Preface

The Best of All Netherworlds is written for both the technophile and the individual who would like to know more about the expanding influence of artificial intelligence (AI) on a society that doesn't quite know how to deal with it. AI is replacing humans with machines, initially for physical tasks but increasingly for reasoning processes. The implications should concern us all.

The mythical country of Bin is not far removed from the growing reality in America. The pages to follow will take the reader on an adventurous journey into this coming world – a world perhaps never to exist, if humans can control the machines, but worthy of contemplation as we move faster and faster into unknown realms of technology.

The story is unusual, even outlandish in parts, and likely to be enigmatic to some readers. But it's based in its entirety on the true workings of computers and logic. It delves into the fascinating concept of human cognition. It weaves discussions of social issues with insights into artificial intelligence, language acquisition, and biochemistry. It is brief but comprehensive, well-researched, and timely in its reflections on modern technology and human inequality. Its main objective is to address the potentially harmful consequences of a foray into poorly understood technologies without an understanding of their impact on humanity.

Table of Contents

Whether to Return

"They were worse off than us. More homeless, more jobless, more drug deaths, more suicides."

"No.." I flashed a dubious glance toward Alan, but turned away when I saw the hint of indignation on his face.

"Bin was a mess, Peter. Jobs were disappearing. Fortunes were being made in the stock market while wages stagnated for a quarter-century."

"That's how it is here!"

"But Bin did something about it."

I had always had a vague sense of the outrageous wealth disparities in the world of our university experience, but most of the harsh realities had happily escaped me. I thought about the two years Alan and I had lived there, devoted to graduate studies, immersed in some esoteric pursuits of cognitive science, secure in our academic appointments while the winds of societal change swirled around us. Change that had transformed Bin into a much different world, about which my more socially conscious friend had enlightened me.

"I was ignorant of all this four years ago." Alan's words made me feel a little less irresponsible. "It's only since I started writing..."

"And ranting. And protesting."

"Right, right." He didn't care for my derisive humor.

Alan never got mad though, as far back as I could remember. His patient demeanor and graceful good looks belied a toughness that had made him somewhat of an athletic standout during our years in school. Our friendly competitions only went my way if I could work a bout of arm wrestling into the agenda.

"Those were good times, talking about brain science with Dr. Babbage."

"I'd like to go back, but it may not be easy."

I contemplated the idea. We were both young and fit, adventurers, risk-takers, perhaps just imprudent enough to enter a country that had imposed strict immigration limits on even those, like us, who had forged meaningful ties through a professional or academic affiliation. Certainly our own lives had changed considerably since then. After our two years in cognitive studies, Alan had turned to social activism, while I had gravitated toward the technological side of cognition, especially with regard to artificial intelligence -- AI, for short -- and the proliferation of robots. I knew that our former host nation had battled with the same overriding social issue that we face now at home: technology's rapid outpacing of humanity's response to change. We don't know how to keep the robots (and their financiers) from controlling our lives. Alan seemed to have recognized this early on. But I was too preoccupied with the microanalysis of human thought -- not an unworthy pursuit in itself -- to step back and comprehend the turmoil around me. Adding to my inattention is that the revolutionary changes taking place in Bin have not taken root in our own country.

It was Alan's activist spirit that ignited the sense of adventure in both of us, stifling any protestations about risking a return to Bin, helping to overcome our anxieties, spreading balm on the pangs of our wanderlust. We had been disappointed in ourselves for being so indecisive. After all, we were recent graduates of that nation's best school. Certainly no harm in reconnecting. We discussed the possibility for several days, alternately coaxing and shaming and cajoling and absolving each other until we agreed that we'd forever regret a decision not to return to the site of our most memorable academic experience. Besides, this could be the ideal summer vacation for us. But it would take much more thinking and planning. We knew little about Bin when we were graduate students four years earlier, and now the politics and technologies were much more complex. We began to take a significantly greater interest in that part of the world that had taught us about minds and machines.

What We Learned about the Nation of Bin

Much had changed, much more than we might have imagined after a few short years. Alan and I had heard the sporadic reports about the developing authoritarianism in Bin, but we only paid sufficient attention when we considered going back. Now we had a much better understanding of the state of affairs that would greet us there. Much of it discouraged an attempt to return. All of it was fascinating to us.

~~~~~

Like a modern-day Atlantis, the nation of Bin beckons with tales of a mystical and seemingly mythical world: pseudo-human intelligence, instilled and cultivated in the technological creations of its people. We have experienced this on a lesser scale in our own country. But Bin is frightening in its rapid metamorphosis to a work-free society. Perhaps this is for the better, though, as their citizens aspire to creative uses of their time -- provided that the decisions made by machine are not to preempt those of its masters. But one can never be sure of that.

The social changes are in large part a response to a dramatic surge in inequality, brought about by the loss of jobs to robots, and by the transfer of jobs to cheaper locations. Bin's productivity had risen sharply for 40 years while wages barely increased. Businesses catered to stockholders. The nation's financial sector found new ways to impose fees and credit costs. In response, and in stark contrast to partisan discord in our own country, liberal and conservative elements of the ruling elite came to agree on guaranteed productivity payments to beleaguered citizens, either to provide basic needs or to get government-run social programs out of the picture.

Much of the technological progress in Bin is manifested in the mundane activities of everyday life. A woman accepts a package delivery at her home, hurries off to class, grabs a taxi downtown, consults with a financial advisor, meets her family for dinner, and then takes the train home. All without being served by a single human being. No delivery person, no teacher, no cab driver, no financial advisor, no food server, no train conductor.

The young woman is one of the few human employees at one of the nation's great manufacturing complexes. Along with several robotic inspectors under her tutelage, she monitors the 4-D printers (time is the 4th dimension) as they schedule and generate massive modular sections of buildings and bridges and solar roads, all to be transported by driverless trucks and assembled by robots.

The woman jokes about going entire days without human contact, for even her meal breaks are catered by courteous automaton servers, and occasional visits from security and the nursing station are wholly machine interactions.

But, rather bizarrely, her emotional needs are being at least partly met by the very machines that have replaced her co-workers. Artificial intelligence and neural network technology is growing up. It allows a robot -- carefully designed to avoid looking too unnervingly human -- to sense an individual's mood, to respond accordingly, to interject its own 'feelings' into a particular situation, to keep learning and maturing to an unimagined level of humanlike behavior. Including the discomforting reality of romantic relationships with machines that are carefully designed to mimic *everything* human.

On a societal and practical level, thousands of people have been kept busy with the company's solar road contract. Bin is crisscrossed by many kilometers of silicon highways that not only power the driverless vehicles and nearby businesses and residences, but also, controversially, provide the sensors and surveillance needed to police the neighborhoods. Outlying roads are lined with nearly motionless vibrating wind turbines. Robots and people have worked side-by-side to construct the free-energy infrastructure, although the human role has steadily diminished.

Yet no one is required to work. Nature's bountiful energy sources have created a society of contented citizens engaged in pursuits of personal interest, some in the arts, others in practical or experimental ventures, and yet others in health care and various humanitarian causes. Some residents serve as informational guides (tutors) for a knowledge-seeking population. Most have time to engage in various forms of recreation. It was certain that such a world would evolve. Work as it was once known in Bin is becoming obsolete. Bin discovered years ago that, perhaps contrary to intuition, a guaranteed income encourages productive family-oriented activities, rarely leading to an overindulgence in the 'temptation' goods. Wealth, of course, exists as never before, but it is becoming less important as self-sufficient communities spring up throughout the nation.

These were the reports surrounding the remote, mountainous region of Bin, and Alan and I remained determined to experience the unfolding realities first-hand. We had an ideal contact person: our old mentor, Dr. Henry Babbage, a linguistics professor and cognitive scientist at the renowned Logic Institute of the progressively high-tech nation. But we were soon to learn that a past connection meant little in the present. Great mysteries lay before us, presumably, in the unknown degree of machine intelligence achieved by the knowledge architects of Bin in just a few years. But the rumors quickly turned wild-eyed, fantastic. Arguments have persisted since Turing's day about the possibility of a humanlike thinking machine, and, to take it a step further, the instilling of "consciousness" in an artificial device -- even though the essence of human consciousness is itself poorly understood.

Yet claims for this higher level of machine understanding had trickled down to the global scientific community from the knowledge creators of Bin. Unsubstantiated claims. But disturbing claims nonetheless, as allegations surfaced of hybrid models of man/machine neural network architectures. Odd, I thought, that we had been so uninformed at the time of our stay. But Bin's AI revolution came about more quickly, and more recently, and with much greater impact than here at home. And of course we hadn't paid sufficient attention to the world outside our immediate concerns.

One other aspect of Binian mystery confronted us: the difficulty of getting back into the country. Despite its social progressiveness, Bin had become increasingly isolationist. As a small, autocratic, naturally fortified outpost to a troubled world, and with a jealous possessiveness of its secretive research, its internal affairs had escaped scrutiny until the reports of technological wonders slipped into the mainstream. Bin had come to reject most forms of collaboration. Foreign students, like we had been, were no longer welcome. More significantly, the increasingly authoritarian leadership had effectively severed the already fragile relationship between our two countries. Bin's national security technology now served as an impediment to outsiders. Military police -- human and robotic -- guarded the borders.

Dr. Babbage, our sole link to the enigmatic nation's technical elite, had participated in an international cognition study several years earlier, in a more cordial period of time, and, enticed by the superior level of research in his adopted locale, he had decided to stay. Alan and I had been two of his research fellows. We blissfully recall spending a final few weeks with him at the university, participating in scholarly post-graduate cogitative sessions in his office on quiet summer evenings, engaged in the free exchange of ideas and truths -- although Alan and I were generally the beneficiaries of any transfer of knowledge.

Unfortunately, subsequent communications between Babbage and the two of us had been infrequent and uninformative. Electronic transmissions to and from the country were largely prohibited. Under current circumstances any form of personal contact seemed unlikely, and any attempt inadvisable. But the good doctor had not ruled out an eventual reunion of sorts, if the political climate would relax its hold on visitor restrictions.

Babbage was engaged in language studies for the summer months at an outlying branch of the Logic Institute, in the rugged foothills not far from the national border. With the youthful bravado and recklessness that often drives adventure-seekers to acts later deemed unwise, Alan and I at last decided to combine our love of climbing with the equally appealing promise of a foray into the mecca of modern technological expertise. We tried to convince ourselves that we'd be accepted as seekers and sharers of high-level knowledge, and as proponents of the need for a common inquiry into the uniting of people and machines. And, of course, as former Binians.

But to reach this futuristic world we would be swept through a tumultuous layer of technological upheaval, with its own dark side to counterbalance the otherwise superb achievements within Bin. As the nation surges toward a way of life driven by machine intelligence, fertile young minds engage in quixotic forms of game-playing and fiendish demonstrations of logic that perplex more conservative members of the fast-changing society. Artificial Intelligence has become the drug of choice for those capable of controlling the machines. For them the real and the virtually real come into conflict as the line between human and machine is blurred.

We wondered aloud if our effort to discover the new world of Bin would prove as futile as the mythological quests for primitive rainforest tribes forever shielded from civilization. And perhaps as perilous. Yet our journey, if in any way successful, was quite in contrast to an expedition to the proverbial heart of darkness, for in this case ours was the more primitive of the two cultures.

Quest I -- Mountain Passage

Our journey started with a treacherous descent
to netherworlds of hopelessness and discontent
amidst the caverns of the Circle First of Hell,
our mortal souls succumbing to the demon's spell.

From a distance, the institute surroundings were breathtaking. Graceful foothills encircled a gentle land of valleys and streams and farmland, exquisitely dark green and fertile in the moist late-summer air. Black forests hugged the peaceful slopes, and brilliant bursts of sunshine seemed to bounce off the waterways as I turned my head to scan the refreshing scene. Far ahead, incongruously at the center of the time-perfected verdancy, stood the milky-walled interlocking segments of a modern research complex. This was human engineering sprung from nature. This was El Dorado. The captivating silence eased me back to happy days of childhood – awakening to the twitter of birds outside my window, a walk through the rows of balsam in a glistening December snow. I felt for a while that I had returned to an earlier time.

"Probably farther away than it looks...c'mon, Peter. Too tired to go on?" Alan seemed amused by my contemplative state.

"Huh? OK. Just taking in the scenery."

"It is beautiful, isn't it? Doesn't seem like the right place for high tech."

For this was a major branch of the Logic Institute, the renowned center of abstract and applied logic; a laboratory of mathematically precise reasoning, headed by proponents of masterful theories of cognition, linking human and machine, explaining behavioral impulses in much the same manner as computer scientists describe the neuron-like pulses racing at nanosecond speed through atomic-level flip-flops. This was the temple of reductionism, where subjective thought processes, no matter how erratic, could be subdivided into discrete chemical and electric charges through which unpredictability ceases to exist. Through which good and evil are quantifiable.

My best friend had always been a thrill-seeker of sorts: a man with passion for social justice and dismissive of personal wealth accumulation, but never unreceptive to life's occasional invitations to bold new experiences. I could tell he was loving every minute of our rather punishing journey through the foothills. He stood waiting impatiently for me now a few feet down the path, amidst the dusty, nettle-choked rock fragments. Before us was a narrow, precipitous ravine that split two foothills and dropped in jagged steps to the valley below. Tired-looking pines, crippled by years of buffeting winds and crumbling earth, leaned across the chasm as if to plead with us to turn back. A most intimidating start to our second day of travel, but a hard-won step toward a pleasantly anticipated reunion with Dr. Babbage.

But the path down the mountainside remained treacherous during the early part of the descent. Apparently we had missed the trail earlier recommended to us. Our progress slowed considerably. Throughout the morning and most of the sweltering afternoon we groped our way down the obstinate slope. At times it appeared to swallow Alan, then spit him up for another few minutes of torturous effort. Slippery gravel, unsteady rocks, and mispositioned boulders hampered us continuously, as did the spiny overgrowth that somehow thrived in the dusty rockbed. A brilliant sun burned our teary eyes and forced us to lower our heads as we stumbled onward. Occasionally, a forward glance would reveal a blurred, rusty panorama that remained annoyingly repetitious. My concern for Alan's condition directed some of the pain from my cramped thighs, but the passing hours proved him a hearty traveler, and in fact similarly concerned with my own level of endurance. By the time we approached the foot of the mountains, we were both more than ready to postpone the remainder of the journey.

"Did you notice the position of the rocks up ahead?" I called to Alan as he dropped his backpack and slumped wearily against a room-sized boulder.

"No, why? They all look just as bad."

"Our path takes us right into that narrow passageway. It might be a dead end."

I peered ahead to the shadowy base of craggy rock that loomed clumsily as our final obstacle. A smooth-walled crevice, obscured by an unnaturally aligned series of grayish boulders, served, seemingly, as the only exit from the ravine. On either side rose ten or twelve feet of sheer rock and, above that, virtually unscalable slopes of sand, roots, and overhanging stone, whose perils might be tested only as a last imprudent resort. Indeed, the darkened crevice provided the only real alternative to retracing our thousands of arduous steps through the mountain pass.

Despite our impaired conditions, we quietly sensed the need to explore the narrow opening before nightfall. We rested for a few minutes in the lengthening shadows of the rocks, and then, after our gear had been secured, Alan followed me down the last gravelly incline to a massive boulder that had likely been straining, inch by inch, century by century, to seal the obscure passageway forever. We timidly entered the area of stone, cool and sun-starved in its gravelike depths, and in fact a welcome relief from the intense heat.

"Do you have a flashlight? My phone's dead."

"I left it in my backpack. Should we go back and get it?"

"Let's go a little farther." The thought of turning around and climbing back through the dust and gravel blunted any impulse toward common sense.

As my eyes slowly adjusted to the shadows, I perked my ear to the muffled emptiness before me. Feeling enwombed by the seductively smooth black walls, I began to step forward, hoping for an expansion of the passageway but increasingly dreading its inevitable end. A thin strip of light from above revealed the interior of ancient stone split by the pressures of convulsing earth. Faintly visible were the orangeish ripples of time etched into its sides by prehistoric waters. The mustiness, the blackish chill rising from the broken floor, and the dark uncertainty ahead of us made me shiver, and reminded me of a long-forgotten moment in a bleak, stale-smelling hotel hallway.

We inched carefully through the damp corridor, testing the cold stone with our outstretched hands as the cave-like passageway twisted one way, then another, in the growing darkness. A mulish determination drove us on. At each step we peered expectantly for a sliver of light that might signify an exit, but only the narrow opening above us, itself gradually succumbing to the diminishing daylight, provided any contact with the outside world. Abruptly, and not unexpectedly, we were halted by a barrier of smooth rock, a strangely well-formed pattern of stone that enclosed us like the three walls of a tall, tiny closet. We groped in the semi-darkness for an opening, but none existed. We would have to turn back.

The chilly underground air, the eerie stillness, and the sensual deprivation combined to hold us in that lifeless corner for a few moments. A vague sense of dread overcame me; I detected a faint scraping noise coming from the walls beside us. The noise continued, gradually revealing itself as a sound like that of sliding stone. Too late we realized what was happening. Alan turned back towards the entrance, but almost immediately encountered a solid wall blocking our exit from the passageway. Helpless, and shocked by the suddenness of the event, we froze before the grunting monolith as it slid the last inch or so into place between us and the entranceway.

I could not comprehend what had taken place. I couldn't think at all, but only sensed Alan's frightened eyes, a few cramped inches away, staring at me through the near-darkness. He muttered a curse, and then began examining the sliding door for a release apparatus. My own slowly clearing mind groped for nonexisting courses of action. Nothing could be done. Four steel-like stone walls, each perhaps fifteen feet high, formed a rectangular shaft around us, and prevented little more than a few short steps one way or the other. A small square of light above remained the only connection to the outside world. Light-flash sensations of confusion were ignited inside me. A flurry of disconnected thoughts dizzied my mind: this must be a security automaton; maybe it was the dark whimsy of engineering students, or even an animal trap; I remembered the sight of a mouse decaying at the bottom of a polished steel pot; I couldn't believe that we had so carelessly allowed ourselves to be trapped, for whatever mischievous or devious purpose. Competing emotions in both of us converged finally into anger at the border security engineers likely responsible for such a reckless and potentially dangerous experiment.

In silence, and in grudging submission, we huddled at the bottom of the dusky pit to wait for our captors to release us. With the return of my senses came the realization that the enclosing wall was fashioned not of stone, but of some type of metal, thin but inpenetrable, as of a steel-sided strongbox. It forced us uncomfortably together, so that little room remained to relax our stiffened, upright forms, and little hope existed for escape. The significance of the latter thought drove a new wave of apprehension through me, and I struggled to remain calm.

The minutes passed. A half-hour went by, with no sounds or signs of rescue. As I waited my eyes became fixed on the top of the shaft. Clearly enough, freedom beckoned us, or taunted us, from the opening above, but its bluish allurement faded into a faraway sky. Steel-slick, fifteen-foot barriers deterred us, as did the unfortunate decision to shed our backpacks and all our supplies before entering the cavern. Inevitably, we decided, only Alan's agility supported by my bulk and strength might free us. In hurried whisperings we planned the attempt, and then, as approaching twilight steadily paled the space above us, we prepared to outmaneuver our unseen captor.

I braced myself between two of the stone walls, and then Alan climbed from my cupped hands onto my shoulders and back to my hands as I gripped his heels and strained to boost him toward the opening. But my efforts nearly failed as Alan inexplicably paused near the top. I felt his weight fluctuating in sporadic shifts of pressure against the palms of my hands, as a cat wavering in indecision before a leap.

"I'm almost at the top," he called back excitedly. "If you can boost me higher..."

Memories of adolescent competitions flashed before my tear-stained eyes as I clutched one of his feet with both hands and pressed upward with all my quivering strength. As volleys of pain rushed through my shoulders, I felt Alan clamp his other foot to the glassy wall and propel himself towards the top. He seemed to hang there for a moment, and then in the next frightful instant his chaotic form, limbs flailing wildly, exploded from the tiny opening into my face and flushed the two of us to the stark black dampness of the stone floor.

For a moment we lay bewildered in a diagonal heap against the walls. Alan carefully unpinned me, then reassured me with a gentle clasp of his hand on my shoulder. "I almost made it," he exclaimed confidently.

His words revived the bit of hope still inside me.

"Are you sure? Can you do it?"

"Damn, I came so close." He seemed oblivious to my words, conscious only of the liberating edge of steel far above that had barely eluded his reach. The determination in his voice convinced me that we had a chance.

We waited a short while, then tried again. Before I was fully recovered from the previous attempt I found myself wedged between the walls, pumping every fiber of muscle into Alan's stretching frame and squinting through sweat-blinded eyes to see him lunge for the opening. He seemed to freeze for an instant. His feet clung to the slippery walls, then suddenly kicked loose and plunged toward me! I recoiled beneath upraised arms, but never felt the blow -- Alan had clutched the lip of the opening with his fingertips! Excitement gripped me as I realized what had happened. I felt an overwhelming desire to help him, but my outstretched hands clenched the air far below his dangling figure. He raised one leg and jammed his foot into the corner to propel himself upward. His laboring muscles flexed into rigidity, as if sculptured from the wall, and I helplessly coaxed and prayed and pleaded until gradually, almost imperceptibly, one of his arms disappeared over the barely visible metal edge.

"Can you make it?" I yelled, begging for a favorable response.

"I think so." The strain in his voice alarmed me, and for a moment I anticipated his exhausted body tumbling back upon me. But he managed to hold on; and after a pause he lurched upward, and then, to my cries of delight, he yelled something at me and then rolled out of view.

He had done it! Excitement pounded at my chest with a drumlike rhythm as I called out to him from the bottom of the shaft. My eyes, my voice, all my energies were directed at the tiny square of bluish-gray light above, which should at any moment have framed Alan's outline as he peered from above to answer my call. But my shouts went unheeded, and I began to sense that something was wrong. Seconds had passed without his response, without his reappearing at the surface. Why didn't he call back to me? He had fallen, perhaps, from the top of the shaft! I became afraid, even moreso than before. I was alone in the lightless pit, perspiration trickling to my lips as I looked up in vain.

But no -- of course! -- he had returned to retrieve our backpacks. That's what he must have yelled at me. The sudden realization calmed me, cheered me, overwhelmed me with inappropriate tinglings of pleasure. All was well now. I waited. I waited many seconds more.

I thought I heard the faint hum of an engine in the distance, but my nervousness distracted me from the sound. Alan still had not returned. Perhaps, it struck me, he had lost his way amidst the great rock formations that encircled us. My cries may have been muffled by the heavy walls. Emotions reeled within me again as my heightened calls reverberated through the shaft.

Abruptly the shadows changed above me. I looked up, and all my nerves seemed to fire at once when I saw Alan at the top, hurriedly lowering a rope to me. Straining to reach it, and experiencing for a panic-filled moment an inability to move at all, I scrambled, finally, frantically, up the walls into the light and towards the growing roar of an approaching motor. Just as I reached the opening, I sprang back as headlights hit the ground in front of me.

"Alan, what..?"

He grabbed my arm and jerked me from the shaft, while shouting at me to hurry after him. An impulse urged me to turn and confront our tormentors. But in the next frenzied moment we were tumbling down a gravel embankment towards the dense, night-shrouded forest, and away from the steady clamor of a vehicle revving its engine some indeterminable distance behind us. We staggered to our feet and tripped through tangled overgrowth to the wooded refuge, never daring to glance back into the gloom, as if in hopes that unseen spirits wouldn't snatch us from the night. We ran together, supporting each other as we stumbled and battled and crawled through invisible obstructions in the thickening wilderness and into the silent embrace of a black pine forest, where we went on and on, deeper and deeper into the insulating woods until finally, far removed from the perils behind us, we could run no more.

Quest II -- Labyrinth

A garden graced the entrance to the Institute,

but some ambiguously troubling attribute

impeded us: a checkerboard of green, festooned

with brambles, barbs, and boundlessness, and souls marooned.

We awoke at that time of dawn when the early grayness yields to a hint of pink and blue, and when imperceptibly brighter passing moments signal the beginning, rather than the end, of day. We were shivering, and we were sore and hungry, but we were happy. The frolicking of the squirrels in the branches above us, the pine-flavored air and sweet earthiness of the forest floor, a nearby creek with a steady trickle of drinking water, and the promise of warmth through the morning mist seemed to revitalize our spirits. We jumped from our flimsy cover of pine boughs and laughed our way through an impromptu wakeup dance that both warmed us and heralded our appreciation for the new day. For we felt we had somehow stolen victory from our adversaries in a bizarre sort of contest.

With the shadows already beginning to melt away from the jagged spruce tops, I felt the need to continue our journey. But our aching muscles and a nagging uncertainty about the previous night prompted a stay on the pine bed to discuss our ordeal.

"I couldn't see when I climbed out," Alan recounted with a shake of his head. "There was sort of a road to the side. I found a spot where I could crawl through the rocks to find the backpacks. I grabbed the rope, and then had to work my way back. I'm glad you kept yelling or I wouldn't have found you."

I felt strangely gratified by the knowledge that my loss of composure had been of assistance.

"I'll tell you," Alan continued, "that truck was getting awfully close. A few more seconds and they would have caught us."

"Caught us? Why? Why a trap out here in the hills, nowhere near the research center?"

"I don't know. Maybe it's a psychotic way to wall off the country."

"Well, it scared me, that's all I know."

Speculation about motives could, of course, be woven to our satisfaction, but the reality of the present -- the prospect of a few more hours of hiking without food or other provisions -- loomed before us like a gathering storm. For in our hurry we had left much of our gear behind. It seemed more productive now to continue toward the university rather than to backtrack through unfamiliar forestland to conduct a dubious search for supplies.

As we began walking, the first silvery rays of sun danced through the misty pine and maple, and scattered to the ground in brilliant jewel-like pieces. A gentle breeze rustled through the leafy poplars in changing waves of green, providing a pleasant contrast to the dark spruce forest behind us. The airy whisperings of the trees disguised the flow of a nearby creek and the scamperings of little animals panicking through our brief intrusion. I envied the spirited creatures their freedom and self-reliance. Yet it occurred to me that they scurried in constant circles, put work before play, and hastened about to avoid being prey to their personal foes. As, perhaps, did we.

Sweet innocent reverie, like a playful child, served to shepherd me from my concerns on that radiant morning. Alan seemed equally entranced by the pastoral setting. But clearheadedness could be of more practical benefit, as I quickly reminded him.

"Alan, didn't you hear me?"

"What? I'm sorry."

"It's too wet over there. And thick. We'll have to head southwest until we find our way out. I think the map showed the university as just west of the river, anyway."

Neither of us was a woodsman, or -- notwithstanding any self-delusions -- a true adventurer, or particularly resourceful in the art of scavenging for food. Our security lay far behind in the bookshelves and computer terminals of paneled dens, where hardships consisted of poor lighting and equipment malfunctions. We stood now in foreign surroundings, bereft of supplies and lacking in the common sense of survival, dependent on memories of questionable mappings to guide us to more familiar terrain.

As we labored through the sticky, dreary-brown underbrush, which sucked and tugged at our ankles with each plodding step, we tired quickly, and rested frequently on the fallen logs that obstructed our way. This was taking much longer than expected. We grew increasingly hungry as the morning hours quickly passed. From the creek that roughly paralleled our southerly route we drank often, while remaining watchful for any recognizable sign of food. The first appearance of unripened fall blackberries stirred remembrances of long-dismissed wilderness lore: the nutritiousness of some inner barks, pine needle tea, and cattail root. The least implausible choice of birch bark proved unpalatable, and so I promptly discarded the idea. Thereafter my hunger was exceeded only by my irritability as Alan led us through the tumorous forest growths on an uncomfortably warm day.

An inner peacefulness long associated with the woods deserted me that afternoon as we experienced a world devoid of the barest expressions of civilization. Exhausted, overheated, and cramped, stinging from the combined assault of insects and thorn bushes, and dizzied by the glare of the afternoon sun, we persisted with a dwindling trust in the travel plan that had promised a research complex but instead had more than doubled our estimated hiking time. On and on we labored, little by little through barriers of stickweed and cockleburrs and low branches and clouds of angry insects, until a destructive pattern had developed: we would grope forward a few steps, then rest; struggle again, then rest. As the afternoon wore on our progress was steadily lessened, our rest periods prolonged. But still we pressed onward. It was with surprising suddenness that we realized, finally, that we stood on a grassy hilltop overlooking a lush valley, where, perhaps just a kilometer away, glowing white in the rising sun and accentuated by the blackened woodland behind it, appeared an interconnected complex of flat cream-colored buildings, looking from a distance like an exquisitely sculptured soap-carving. We had reached the Logic Institute.

We paused to take in the scene, a confounding mix of old and new, of expected and extraordinary. Just beyond the modern structures, amidst graceful willow trees and thick barriers of pine, stood a marvel of Gothic architecture, a relic from a long-forgotten age, a towering greystone masterwork of arches and ornamented buttresses reminiscent of Renaissance cathedrals. An expansive grassy terrain, decorated with hedgerows and dark patches of dirt in a labyrinthine crisscross pattern, stretched along the entire length of the new and old sculptures. A river lay just beyond that. The entire grounds sat peaceably amidst fertile green meadows that dipped and swayed for miles in every direction. I was struck by the stillness of the landscape. This was an idyllic location, ideal, apparently, for the purpose of minimizing outside distractions for the young engineering and cognitive science students. It seemed odd that no people or vehicles could be seen on the premises.

I stood amused by the contrast between new buildings and old. It seemed too unreal, too contrived, like a child's building-block city that rises and falls in frivolous bursts of energy and serves no purpose other than for the amusement of its little creator. Perhaps, I thought, my critical first response was due to an ill temper built up through a wearisome journey.

Hesitantly we descended from the grassy hilltop, toward the extravagant patchwork of garden that sheltered us from view of the buildings. In a matter of minutes we were standing before a massive bronze arch adorned at both sides with dense rose bushes, a neatly trimmed lawn lined with marigolds, and a two-sided hedge wall that appeared to extend some distance before us in the manner of a garden pathway, the breadth of which approximated that of my extended arms. From where we stood, we could not yet see the complex itself.

We stepped through the imposing arch and started along the

flowery lane, walking perhaps thirty meters between two rows of dense, prickly hedges before approaching a forked pathway and a neatly inscribed sign directing us through one of two bronze gates towards the complex:

IF NOT A STAFF MEMBER OR A VISITOR
PROCEED TO THE RIGHT, ELSE PROCEED TO THE LEFT.
(*Auth'd by Prof. Boole*)

We were visitors of a sort, so after a moment's hesitation we proceeded through the left gate.

"That's confusing." Alan looked perturbed as we turned into a double row of hedges that extended well above our heads. "Why couldn't they just say 'staff and visitors proceed to the left'?"

"I know," I agreed with him. "I remember Boole. He's a Logic Institute dean who never seemed that logical. Something funny about this sign."

To the purpose of avoiding deterrence in any form we were committed, having pledged ourselves to watchfulness since our ordeal in the mountains. We felt certain that once inside the

complex Dr. Babbage and his students would provide the
necessary sanctuary.

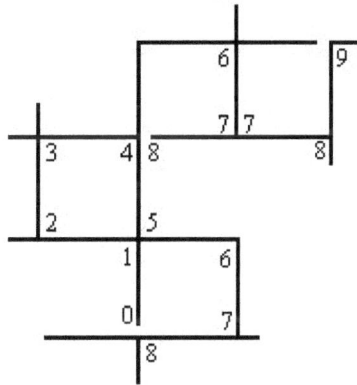

After entering the garden we quickly encountered a T-shaped
branchpoint in the path, its right fork labeled with a sign stating
"PATH TO MAIN ENTRANCE" and indicating the distance as .8
kilometers, about a half-mile according to my hurried calculations.
The double row of dense green hedges guided us to another sign
denoting a distance, to the left, of .7 kilometers. We followed a
mandatory left turn at the next appropriately marked turnoff, and
then proceeded to a crosspath of sorts, the intersecting point of 4
identical pathways, where we turned right at a sign labeled as .5
kilometers from the entrance. There we entered a straight pathway
that was interrupted about half-way down by a .4 kilometer sign
pointing left. By this time our interest in the well-groomed
surroundings had deteriorated into a simple desire to reach our

destination. We hurried along the narrow pathway: .4 kilometers left, .3 kilometers left, each football field length sapping more of our strength, each ninety-degree iteration imposing another mirror-like reflection on our confused minds. Alan was the first to detect a hint of misdirection in the pattern, and his suspicions sent a flash of trepidation through me. Yet I quickly dismissed the thought as inconsequential, in view of our apparent proximity to the main entrance.

We were directed .2 kilometers to the left, and then .1 kilometers to the right at a 4-way crosspath, but each change revealed nothing more than a continuation of the carefully manicured garden walls. We quickened our pace to reach the next junction in the path, anxiously anticipating a glimpse of the logic building. But we encountered only a cul-de-sac of thick shrubbery that provided no outlet. We had no alternative but to return in the direction from which we came. I stood motionless for a moment, puzzled and disoriented, a familiar uneasiness clutching at me from inside. Alan advanced to the end of the blocked pathway and called out to me. He stared blankly at a sign that described the "PATH TO MAIN ENTRANCE" as .0 kilometers removed from the point at which we stood.

"This is ridiculous. What's going on here?" Growing irritation marked Alan's words, and a feeling of helplessness overtook both of us.

We attempted to estimate our position relative to our starting point just inside the bronze gate – an unenviable task at a position .0 kilometers removed from the point at which we wished to be; and we decided, after brief deliberation, to return to the previous kilometer markers in an effort to reconstruct our path. After retracing our steps to the last junction, we turned left, verified the presence of the .1 kilometer sign, and continued to the next crosspath, where we located the .2 kilometer marker and proceeded to the right. The seemingly unpatterned sequence confounded me, but Alan recalled two left turns (I thought it was three, but Alan was certain it was two) before the .1 kilometer crosspath, which reversed to two right turns upon our renegotiation of the route. Sure enough, a second right turn guided us to a T-junction and the .4 kilometer sign, where a left turn led us eventually to an unmarked right bend in the path, the discovery of which cheered us as we recalled an obligatory left turn at an early stage of our inward route. Some distance beyond this point we turned right at a crosspath, confirmed the presence of a .6 kilometer marker, and then advanced towards a T-shaped branchpoint that momentarily halted our progress, for both paths leading into it pointed .7 kilometers in the direction from which we came, the alleged route to the entrance. Fortunately, we recalled beginning our wanderings at the .8 kilometer sign with a right turn followed by a left; and so, despite some momentary recalcitrance on my part, we reversed direction and proceeded to the left in anticipation of a subsequent right. As, indeed, we approached another T-shaped branch in the path, I began to relax in expectation of the gated entranceway preceding the .8 kilometer sign that had initiated this futile chase. We turned right, but rather than the expected gate, we faced another confining column of hedges that appeared ominously black-green and impassable in the fading afternoon light. Together we stepped back to the signpost. Appropriately, it announced a .8 kilometer distance to the entrance, but no gateway was evident anywhere along the pathway. Disorientation possessed us, stunned us with its swift return. The path ahead appeared strikingly similar to all the others.

"We did turn right at the first sign, didn't we?" Alan's words seemed somehow meaningless after so many right and left movements.

"Yes, I'm sure we did. But if we came in right to left, we shouldn't have gone back left to right."

Alan admitted I may have been on to something there, although by now the proper execution of the lengthy sequence of turns was beyond our comprehension. Nevertheless, we returned to the previous .7 kilometer T-junction and passed straight through, thus reversing our previous left turn to a right. One-tenth kilometer beyond stood another .8 kilometer signpost facing a wall of shrubbery with no apparent exit, entrance, or passageway. Alan's initial blank-faced response to this continued misguidance exploded into a cursing, blusterous return to the previous .8 kilometer sign, where he, and I behind, unhesitatingly proceeded to the left toward the next turnoff. The discovery there of a .9 kilometer marker forced us to the ground, dazed by the suddenness of this unexpected turn of events.

I sat in the grassy path and closed my eyes to the unbroken line of surrounding hedges, so bluish and cool, shadowy and forbidding, and silent in a way that seemed to magnify our impotence. It was definitely getting darker. The idea of being trapped in a garden almost drove me to laughter, but Alan seemed unwilling to share in the mirth. Once again we had allowed ourselves to be ensnared in the instruments of wily gameplayers. My mind was racing. My moods fluctuated like gusts of wind in a winter storm. Perhaps we hadn't stumbled into a trap after all. Instead we had simply turned the wrong way, and that was all. We should wait for them to come to us.

For a minute or two Alan and I sat in an awkward silence, experiencing the volatile moods of roulette players with stacks of money on the table. The cool evening breeze returned me to a recent outing with friends who were seeing me off to school. I tried to visualize them behind tightly closed eyes. Slowly – too slowly, it seemed – the image was reconstructed in the jumbled patterns of light within my mind. And then, as in a misshapen carnival mirror the image was stretched and distorted into a collection of grinning specters that lured me away from the realities about me.

Then I heard a voice.

"Alan, did you hear that? From the hedge over there." I turned in the direction of the soothing monotone sound, under the same quizzical spell that seemed to transfix my friend for a few short moments.

HELLO ALAN. HELLO PETER.

"Who are you?" My voice sounded muted within the cumbrous hedge walls.

I AM HERE TO WELCOME YOU.

"How do you know our names?"

The silence that followed seemed to mock my question. I turned to Alan and half-whispered, "Facial recognition...that's a machine talking to us. I'm going to try something." Turning back, I announced authoritatively: "We would like to see Dr. Babbage...Henry Babbage!"

A few more seconds of silence followed -- a disturbing silence, filling my mind with speculation about the sincerity of the 'welcoming' message.

PLEASE WAIT.

I was encouraged by the subtle hint of positivity from our robotic greeter. Alan wasn't. He directed me away from the automaton and spoke in hushed tones: "We should have known there'd be surveillance everywhere. They could be coming to arrest us."

"Well, we're stuck here anyway."

It occurred to me that our customary roles had reversed, with my usually upbeat friend anticipating the worst while I looked forward to the next machine interaction. It came soon enough.

HENRY BABBAGE HAS BEEN NOTIFIED.

"Thank you!" My spirited response was wholly unnecessary.

Alan and I stood on the pathway, too nervous to sit, our backs against the hedge opposite the estimated position of our invisible host. And so we waited. It was late afternoon, close to dusk. We were both thirsty and hungry, and we looked forward to some sustenance and assistance from our old mentor.

"Alan? Peter?" The suddenness of Babbage's greeting sprung us to our feet.

"Henry, it's great to hear your voice!"

"You shouldn't have come." His tone struck me as all wrong for this long-awaited moment.

"We wanted to see you...and find out about our research projects."

"The time isn't right for a visit...it's summer break...new AI...the surveillers might take an interest in you."

Despite the dismissal in his words it was somehow reassuring to hear Dr. Babbage's voice again. A hint of empathy for our situation was unmistakable in his tone. Yet so was his certainty about turning us away.

"Is there any possibility of a brief visit? We need food and water, camping supplies."

"An L.I. assistant will bring whatever you need."

Logic Institute, he meant. The masters of Bin were apparently dead serious about the "no visitor" policy, and Alan and I were being offered a graceful exit.

Babbage continued: "If things change, I'll contact you. Maybe soon.."

Neither Alan nor I knew how to respond to the wistful note in our professor's leave-taking. I groped for a few words: "I'm sorry to hear that, Henry."

"I wish I could say more. Ask the assistant for the garden exit."

With that he was gone.

The two of us lay on the pathway near the .9 kilometer marker, waiting for the promised Logic Institute representative. I pondered the fast-evolving turn of events that had now changed our plans. The brief communication had been puzzling, atypical of conversations with Babbage that I well remembered from years past. Our relationship with the professor had always been more of a sons-to-father connection than that of students-to-teacher. Alan especially. Babbage, like so many others, was drawn to my friend's outgoing personality.

A few minutes passed. Neither of us was ready to return home. We lay there discussing the absurdity of an aborted journey to the nerve center of high-tech, after three days of travel and an inexplicable rejection by an intimately-connected educator and friend. Alan muttered at one point that we should re-initiate communications with the hedges and simply try again. But there seemed to be no alternative to our designated departure. At least we would get the much-needed supplies for the re-entry into our own world.

Quest III -- Logic Trip

"Identify yourself," I cried, "and don't forsake
a kindred spirit. Have you come to be my guide?"
The specter talked to me in soothing tones: "We'll take
another path to journey to the mountainside."

Our denouement, a disconcerting cul-de-sac
quixotically devised by some demoniac
or deviant, produced the same distinctive rage
that builds to madness from the corners of a cage.

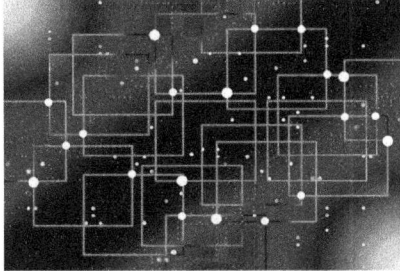

"Peter!"

Alan was staring at me, his face an unlikely picture of composure.

"What is it?"

"The assistant is coming."

I strained to listen in the crisp, motionless air. It occurred to me that Alan may have imagined the sounds of somebody approaching. I turned away at the thought and contemplated instead the gradual blending of the hedgetops with the darkening sky.

A minute or two passed without further sound. I shivered as I peered into the gloomy stillness of the pathway. A sudden flickering of light whirled me around – something had moved. Perhaps it was my imagination. Maybe a rabbit, or a bird. We hadn't seen any wildlife, though. That had seemed strange.

And then we both heard the voice. It penetrated the still night air like a rumbling of thunder, its eerie masculine tone suggestive of the low-pitched reverberation of an echo chamber. The few words spoken were unclear, as if muffled by thickly insulated walls, which in fact was the case. I struggled to make sense of the imprecise words:

I AND YOU DIE – WE DESTRUCT

The message sent a shiver through me, and I nervously turned to Alan, but received only a blank response in return. The ominous words sounded again inside of me. How could this happen after Dr. Babbage's promise of assistance? We waited in stupefied silence in the quiet of the hedgerows. Night was nearly upon us. Words formed on my lips, but they went unspoken as we were disrupted by a booming voice that startled us with its intensity:

I AM YOUR GUY – PLEASE INSTRUCT

"I am your guy; please instruct?"

"I am your guide! Please instruct!" Alan laughed in relief at recognition of the obvious message. In the same instant his silhouette appeared in a backdrop of dim light from a nearby opening in the hedge wall. With a degree of vigilance, and hesitance, we approached the crosspath and prepared to confront the person or object that apparently offered service to displaced visitors. Together we glimpsed around the corner and then drew back to the protective darkness of the hedge walls; for towering above us was a magnificent device of glimmering steel and chrome, its size and box-like shape unmistakably defined but its detail obscured by a powerful searchlight radiating from its center. We froze in wonder at the sight, uncertain whether to advance or retreat, rather like puppies first encountering a great floor-sweeping machine. Seconds passed without movement, without a sign of life from its monstrous bulk. Clearly we faced an electromechanical vehicle of some sort, equipped with sensing devices for the negotiation of the pathway angles and a speech capability that so far was limited to the repetition of its initial announcement. We waited for a reiteration of the message, but the silence persisted. I surmised that our presence had temporarily deactivated the metallic giant. The hypnotic aura of the moment – the yellowish cone of light casting spectral shadows against the bushes, the embalming serenity, the blackness above, behind, and beside us – drew us closer to our finder's grasp, to be subjected to its whim or official duty.

"Who are you?" My query sounded a bit more aggressive than intended.

"I AM YOUR GUIDE. PLEASE INSTRUCT." The bass-like, monotone interruption forced me to retreat to the edge of the bush and focus inquiring eyes on the metal outline of the machine.

"Where is the garden exit?"

Again a short pause preceded the throaty cadence of its mechanical words. "PLEASE INSTRUCT."

"Take us to the exit."

PLEASE INSTRUCT.

Exasperation glowed with a hint of pink on Alan's face. He turned to me as if for reinforcement, and after a moment's thought I obliged.

"How do we instruct you?"

PLEASE ENTER THROUGH DOOR.

We cautiously stepped to the side of the apparatus, where, less directly hampered by the forward-searching beam, we detected a sizable opening in its metal frame. Evidently the peculiar structure, which resembled a narrow storage shed – although somehow more gracefully designed – could accommodate two or more people for the purpose of its intended instruction. With my perfunctory urging Alan forced his way between the smooth-sided device and the brambly hedge wall to the weakly illuminated oval doorway. I followed. Visible inside were a pair of straight-back chairs with overhead harnesses, apparently intended to be lowered over their seated occupants. A communications device glowed from the front wall of the little compartment, facing the chairs. Most welcome to our adjusting eyes was a tray with sandwiches and water bottles and sweaters and blankets and traveling gear.

Alan finally looked a bit cheery to me. "Just in time," he remarked. It's getting dark. And cold."

As we ate I examined the communications screen in front of us. It was blank except for the soothingly suggestive phrase 'Touring Machine' in an upper corner. I stood immobile for awhile, alternating glances between the screen and Alan's barely illuminated face, which glistened top to bottom with beads of perspiration like that of a man tending a furnace rather than the cool interior of a robotized vehicle. At the moment I began again to speak, an abrupt blinking from the screen froze me in place and lured Alan from his position behind the food table. Fully engaged in the oddity of the moment, we stood gawking like boys at a magic act as the interactive device finally resumed our suspended conversation.

GARDEN EXIT OR LOGIC INSTITUTE ENTRANCE?

We immediately turned to each other. We still had a choice. The need to turn back was suddenly less certain, and a thrilling twinge of defiance struck us at the same time. Alan nodded at me, and I uttered the words:

"Take us to the entrance!"

The device's soundless reply was a cryptic set of choices on the screen in front of us:

ANNOUNCE YOUR CHOICES BY NUMBER,
   IN THE ORDER DESIRED.

1 STEP FORWARD.
2 TURN RIGHT.
3 TURN LEFT.
4 IF BARRIER IN FRONT, TURN RIGHT.
5 IF BARRIER IN FRONT, TURN LEFT.
6 REPEAT FROM BEGINNING.
7 STOP.

IF YOU'RE LEFT HERE YOU'LL BE RIGHT THERE.

We gazed in puzzlement for some time, fully expecting additional information to suddenly appear, either by voice or text. None, unfortunately, did, and we gradually recognized the need to coax an interpretation from the available context.

"It's a test of some type," Alan whispered. "They're messing with us again. To get to the school we have to maneuver our way through the garden by choosing the right instructions." He was wholly absorbed now, forgetful of Dr. Babbage's admonition. "We announce them in the proper order to this....machine, and it uses the pattern to guide us out."

"But we don't know which way to go!"

"We don't know which way to go, which way to turn, or whether to turn at all." He was looking at the lighted crosspath just ahead of us, which, of course, presented four directional options.

I was quickly being drawn in to the challenge. My eyes focused on the quizzical statement near the bottom of the screen:

IF YOU'RE LEFT HERE YOU'LL BE RIGHT THERE.

"That makes no sense at all," I said, partly to Alan, and partly in hopes the game-playing apparatus would offer some explanation.

Alan studied the screen intently for a minute or more. "I have an idea," he finally announced, softly but excitedly. "Maybe 'left here' means 'left'" -- he extended his left arm – "in a directional sense. Maybe it's telling us to keep to the left."

I skeptically mulled over the suggestion, but Alan hastily continued. "Here, I'll try it. I'll tell it to step forward and turn left."

"Then it'll be facing the hedges."

"Well, it depends on the length of a step. Maybe this thing will step to the next opening, then turn left."

My better judgment notwithstanding, and only after my insistence on a 'stop' command following his proposed sequence, Alan convinced me of the sensibleness of testing the two movements 'step forward' and 'turn left,' in that order. He turned to the communications screen and leaned forward to call out his instructions.

"One. Three. Seven."

He withdrew at once, whispering his misgivings about the

machine's capacity to accept his commands. Indeed, proper procedures hadn't been clearly defined. After experiencing no immediate response from the device I began to wonder about the feasibility of another attempt.

Before I could react, however, the door suddenly slid shut and the apparatus began to move. Smoothly, as if suspended in air instead of on wheels, the entire unit eased forward about one meter, then swung ninety degrees to the left and stopped. Alan immediately reached for the door and slid it open, to the relief of both of us, for despite the apparent cooperativeness of the device we still feared the prospect of confinement. Yet we gave no thought to leaving the machine. Encouraged by its ready compliance to our commands, we quickly plotted the next directive, which consisted, simply enough, of a right turn and a stop. Rather excitedly I leaned forward to call numbers two and seven, again feeling a trifle foolish in the process. A slight hesitation followed, and then the machine spun suddenly to the right and stopped, facing forward, just ahead of its original position.

"That's great. We've got it now." Alan's exuberance disguised the fact that we had progressed only one meter since taking the controls. "Let's try this," he continued. "Step forward until we reach a barrier, then turn left. That'll guarantee continuous left turns."

Something about his suggestion disturbed me, but the thrill of success had drawn me into the spirit of the moment, and I readily assented to the bold attempt. Quickly but carefully we agreed upon the specific steps: "step forward"; "if a barrier in front, turn left"; "repeat from beginning." Thus, we reasoned, the device would progress without deviation until reaching a left branch in the path, at which point it would veer in that direction and continue unhampered to the next leftward junction, where after another turn the next cycle would be initiated, and so on, and on, and on.

Any doubts I might have had about the sequence deferred to Alan's eager recitation of the new instructions to the communications device. As before, a short delay preceded the machine's movement, and then a sudden, smooth acceleration propelled us forward. The continuous motion caused an uneasy fluttering in my stomach. I realized that the device was indeed obeying our commands, or at least proceeding in such a manner as to appear obedient. Steadily it gained in speed, passing a four-way crosspath without hesitation, and soon hurtling forward with a chest-tightening velocity that threatened to annihilate the first prickly-hedged barrier in its way. From our half-standing, half-stooped positions in front of a small window we could view the grassy pathway whisking beneath the brilliant searchlight, and we squinted ahead in anticipation of the inevitable path-blocking hedge wall. Without thinking I pulled Alan towards the straight-back metal chairs where we hurriedly lowered the leather harnesses to our chests and, while still watching the path walls speeding by, braced ourselves for the collision.

The indescribable pressure that forced my upper body against the harness was all that could be recalled of our near encounter with the hedge barrier. Somehow, almost magically, the vehicle managed to slow itself just inches from the wall. Then it turned left and embarked upon a new pathway. Cold sweat, I noticed at the moment we turned, had saturated my shirt and caused it to stick to my skin. I managed to relax a little, feeling minimally secure in the supporting hold of the harness. But just as quickly I cringed at a new thought, one that embarrassed me with its simple inevitability: what should occur in the event of a right bend in the path? We had instructed the vehicle to turn left if unable to continue its forward progress, an obvious impossibility in the case of an ordinary right turn. We had failed to consider that eventuality. Still, the thought occurred to me that perhaps only left turns remained, and that we may have inadvertently issued the proper instructions for transport to the Logic Institute entrance. The answer came soon enough. Before I could reflect further on our condition, the vehicle swung left and ground to a complete halt before a solid wall of the everpresent, confining hedges.

Alan and I simultaneously raised the harnesses above our heads and scrambled through the sliding door to investigate our surroundings. As I had anticipated, we had stalled at a right bend in the path, at which the lumbering conveyance, after turning left directly into a wall of shrubbery, had somehow sensed its inability to proceed. Instead, its searchlight, reflecting off the glossy leaves to its front and right, revealed the obscure beginnings of the pathway behind it that should have continued our course after a right turn. We stared at each other in a spirit of discouraged, yet still hopeful, embarrassment.

"It's my mistake," I conceded. "I realized too late that this might happen. We're just lucky that the machine stopped instead of trying to go on." The thought of our amorphous guide forging a path through the hedges amused me to the scant degree permitted under the circumstances.

"What now?" Alan's voice sounded high-pitched in the chilled night air. I proposed that we revise the instruction list once more to accommodate unconditional right branches. A few minutes of careful deliberation and note-taking provided, after considerable disagreement, a sensible result:

1 STEP FORWARD.
5 IF BARRIER IN FRONT, TURN LEFT.
4 IF BARRIER IN FRONT, TURN RIGHT.
4 IF BARRIER IN FRONT, TURN RIGHT.
6 REPEAT FROM BEGINNING

Thus, according to our reasoning, the path-seeking device would turn ninety degrees left upon encountering a hedge wall. If still facing a barrier at this point, it would undergo a complete one-hundred-eighty degree reversal through two right turns, thereby realizing the net effect of a single ninety-degree right turn at any right bend in the path. Satisfied that our oversight had been remedied, we turned to the communications screen, ordered two right turns and a stop to reposition ourselves, announced the adjusted series of numbers for our continued progress, and secured ourselves in the protective embrace of the leather harnesses.

Once again the unit began its steady acceleration upon a newly-lighted path which, we hoped, would provide an uninterrupted route to our long-sought destination. Feeling reasonably confident of our latest instructional revision, we settled back as comfortably as possible to observe the upcoming sequence of turns. Before attaining even a medium speed we received a partial affirmation of our accuracy as the device cleanly followed a left bend and re-initiated its forward motion. Just moments later, it repeated the smooth movement at a right bend. I gazed confidently ahead, pondering the simplicity of computer logic, when another quickly-approaching barrier of shrubbery appeared suddenly in front of us. An instant of panic forced my eyes shut as I sensed the inevitable occurrence of a dead end. I braced myself for impact, but in a moment it was over. The device had stalled again, this time facing the right side wall of a cul-de-sac.

Alan and I together sputtered exclamations of relief, then laboriously pulled ourselves from the harnesses to begin the compulsory examination of our position. Both of us sensed the solution to our latest miscalculation. Within seconds we had agreed upon the proper sequence of steps:

1 STEP FORWARD.
5 IF BARRIER IN FRONT, TURN LEFT.
4 IF BARRIER IN FRONT, TURN RIGHT.
4 IF BARRIER IN FRONT, TURN RIGHT.
4 IF BARRIER IN FRONT, TURN RIGHT.
6 REPEAT FROM BEGINNING

"So now," summarized Alan, "we're guaranteed some movement, because if there's nowhere to go, three right turns after a left turn will simply take us back the way we came."

He seemed positive, but a stirring of doubt caused me to question the new sequence. "We're instructing it to move forward as far as possible, try turning left, and if that fails turn right until it can move forward again – so we're guaranteed to continue some form of movement indefinitely."

"Right."

"But how do we stop? And couldn't we get stuck in a circle somewhere, in an indefinite looping pattern of some type?"

"I would hope this contraption stops when it reaches the end. And I don't know about a loop. It seems to me that following the left wall all the way just might get us out of here."

I felt some intuitive agreement with his reasoning, although certain doubts remained. One concern immediately came to mind.

"Did you notice," I asked, "the reaction of this thing at the right turn? It didn't even try to turn left before it turned right."

"I know. I thought it was ignoring our instructions at first."

"Maybe it was using some kind of look-ahead technique."

"You mean there's no need to turn left if the next movement will be a right turn."

"Exactly."

"I've been wondering about the movement of this machine," Alan said. "It doesn't feel like we're riding on wheels. Maybe it's powered electromagnetically, lifting us off the ground as it accelerates."

"I don't know. Maybe." I was eager to continue our passage through the interminable aisles. At Alan's insistence, we tried once more to converse in words with our single-minded guide, but received only the dullish glare of the communications screen as a response. So we ordered the vehicle to turn right, and then carefully announced the adjusted sequence of instructions.

The now-familiar ritual followed in rapid fashion: our securement within the harnesses; the barely perceptible initiation of movement;

the airy acceleration through the tunnel-like hedge borders. Abruptly we slowed for a left turn, then soon after for a right. Neither appeared disruptive to our smoothly navigating guide. Like a barn swallow swooping through familiar rafters, it flowed effortlessly through the angled corners of the labyrinth in its effort to complete its task. In quick succession we turned left, then right, then right again, each time apparently building to a greater speed, until finally the powerful vehicle whipped us ahead in a blur of silvery light through the rail-like, black-green rows of hedges. As the acceleration decreased, I attempted to mutter some encouragement toward Alan, but before the words were out a renewed pressure suddenly forced my chest and right arm against the harness with an intensity that smothered all breath and sound for a frightening instant. Almost immediately the pressure began to subside. After a brief respite it returned, less severely than before but with enough force to tighten my muscles in a conditioned response. Again the pressure eased. And again – before my breath was fully restored – it returned. Less than a second it would last, clamping my powerless upper body against the bulky harness before releasing me for another few seconds of disorientation. Quickly, without interruption, the alternating effects continued: heavy, vicelike pressure against my right arm; then weightlessness, and a release from the pressure; more pressure; release; pressure; release; pressure; release. From somewhere beneath the disfunctioning muddle in my mind came the realization, so suddenly obvious, that we were turning left with a periodic frequency suggestive of a squarelike pattern. For some minutes it continued in this fashion – twenty, thirty, a hundred times – until my right side felt wholly in the grip of a massive fist alternately clenching and relaxing, clutching and releasing, and gradually stealing my breath with its abusive regularity. Finally, impulsively, I began to yell, more from desperation than design, but with the flickering hope that perhaps the senseless device would respond to a blatant plea for mercy. Over and over I called for a halt to the machine's frenzied looping. Before long I was joined by Alan, whose plaintive cries sounded curiously distant in the tiny compartment. Our efforts appeared futile as the vehicle, persisting like the plaything of a child, sped aimlessly along its ill-programmed route. A sudden impulse prompted me to call out "number seven," and upon doing so I felt a streak of pain shooting through my body from beneath the oppressive weight of the harness on my chest. And then I was whipped back against the chair like a sack of grain against a wooden pallet. My trembling, dizzied body lay limp and barely conscious after the unexpected thrashing, and quite unaware of the thorough motionlessness of our eccentric carrier.

Alan's voice penetrated the silence. "You stopped it, Peter. Peter, are you OK?"

I opened stinging eyes upon the familiar scene, through our little front window, of a lighted pathway stretching to the shadowy limits of the vehicle's front searchlight. Painfully, I released myself once more from the harness and sat facing Alan in the soundless gloom.

"I'm OK. What happened?"

Alan's failure to respond attested to his equally confused state. Our last series of instructions had seemed flawless to him, yet our efforts had ultimately failed. We could only stare into the unchanged communications screen and wonder about the wisdom of a further attempt, an attempt that might serve only to satisfy the urges of an inventive fiend who might be observing us at this very moment. I waited in silence, unwilling to speak, unprepared to act, and angered by the repeated frustrations.

But I wasn't about to give up. Sliding the door open, I pulled Alan outside, then crouched beneath the searchlight and sketched a crude diagram of our repetitive pathway in the dirt next to the hedges.

"I was thinking," I half-whispered to Alan, "that maybe we missed something. Look, we entered a square path with all left turns and no exit. It must look something like this." I showed him my simple drawing.

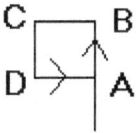

```
C ┌──┐ B
  │  ↑
D └→ │ A
     │
```

"You see, according to our instructions, we got trapped because we went past point 'A' to point 'B' and turned left; after passing 'C' and 'D' we returned to 'A,' where we were forced to turn left and start the cycle all over again!"

Alan considered the picture for a few seconds before peering back at me through the grayish light. "But then the only way out is by a right turn. It seems we'd be trapped just as easily if we requested all right turns."

"No, you don't understand." But with my next words the understanding struck him, as if through a transfer of thought. "We should have turned left at point 'A,' then passed through 'D,' 'C,' and 'B' before returning through 'A.'"

Alan looked on encouragingly as I hurriedly revised my graphic:

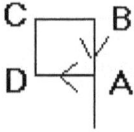

We had recognized the disabling flaw in our strategy, one which had in all likelihood stirred the riotous whirlpool of motion just past. And I sensed a correction that might finally guide us to our destination. The endless loop had evidently been caused by our decision to proceed to a barrier before turning left. If instead we were to turn left at the first opportunity, we'd effectively trace a clockwise path around the outermost edge of the loop, however large or small it may be. Such reasoning should apply to a continuous loop of any size, including the entire garden itself, if indeed it existed in such a pointless state.

But rather than while the time immersed in theory, I quickly formulated a new list of commands for our waiting conveyance, a near-repetition of the previous series:

1 STEP FORWARD.
3 TURN LEFT.
4 IF BARRIER IN FRONT, TURN RIGHT.
4 IF BARRIER IN FRONT, TURN RIGHT.
4 IF BARRIER IN FRONT, TURN RIGHT.
6 REPEAT FROM BEGINNING

"So it looks to its left after every step instead of waiting for a barrier?"

"Yes." I felt fully confident of my decision. "Now it'll turn left at every opportunity – or at least look to its left. If there's no opening to the left, it'll turn back to the right until it can move forward."

Alan instructed the machine to reverse itself with two right turns. He proceeded to recite the revised command list, and in another moment we were whisking through the hedge columns, apparently free of the troublesome looping path, and cautiously confident of our chances for success.

As our transport continued unimpeded, I felt myself surrendering to the gentle fluctuations of smooth left turns and forward accelerations. We were totally at the mercy of the machine now, as it was of our instructions. Great distances collapsed into narrow blurs of motion at our window, and rapid-succession directional changes seemed less and less noticeable as they blended into a singular sensation of flesh-pleasing motion. Within moments I felt I was moving not at all, but adrift upon a rhythmic wave of sweet passion that alternately rose and fell with the seductive stirrings of our carrier.

What I Dreamed

Beyond the caverns of the Circle First of Hell,

where vagabonds and castaways of Limbo dwell,

we stood in Circle Two of Lucifer's domain

and viewed an image of the otherworldly reign

of warrior and governor and financier --

the first with gold accoutrement on bloodied spear,

the second steeped in camouflage and blasphemy,

the third a cunning crafter of complicity.

A mix of memories from my two worlds came to me during the night, in ways either embedded or invented in the quivery corridors of my mind.

As part of a top university's commencement, a billionaire received an award for his "Outstanding Contributions to Society." The commencement program addressed him: "..Your influence is felt throughout broad sectors of society..your commitment to..a better quality of life for people across the world." Eight years earlier he had collaborated with a financial firm to create packages of high-risk subprime mortgages, so that in anticipation of a housing crash he could bet against the sure-to-fail financial instruments. His successful bet against his nation's households ultimately paid him nearly four billion dollars.

In Circle Three, amidst the acrid stench of Lust
and Gluttony, the grinning serpent judge decreed
that we should meet the lords of wealth, and we were thrust
inside a paupers' den, where we were made to plead
for crumbs from unrepentant men of avarice
along the scum-encrusted banks of the abyss.

The leading manufacturer of a children's allergy medication made small changes in the product, enough to give it patent protection, and then began to offer the medication free of charge to schools, gradually making school nurses at least partly dependent on it, while at the same time successfully lobbying for a law that would require the presence of the medicine in schools. Then, after quietly raising the price six-fold, the company announced a half-price coupon.

In Circle Four an image of the "poor and mean
condition" of the masses sent a ghostly scene
around the world as military men embraced
a notion quaint: by seeking peace they died disgraced.

One of the world's richest nations is bombing one of the poorest, using the weapons and technology of the world's most powerful country. Electric and water facilities have been destroyed, supply lines have been cut, schools and weddings and funeral processions and hospitals have been bombed, and a cholera epidemic rages out of control. A bombing victim described the ever-present sense of terror from the skies: "I can't sleep...when the drones are there...I hear them making that sound, that noise. The drones are all over my brain."

In Circle Five are men with scornful grimaces,
all well-to-do, respectable, and dignified,
regaling nightly at their gated premises,
condemning urchins lying hungry just outside.

A woman camps out amidst the stench of a field of trash, not far from sidewalks that are bleached every morning to prevent the spread of hepatitis. Nearby, two homeless people sleep in a car in a superstore parking lot, which to one of them "feels like a palace" compared to the places she's been. A young man has given up hope of begging for the $20 needed for a room for the night, and despite near-zero temperatures curls up in a dark corner under the viaduct. Another homeless woman rides buses all night, but told a reporter, "If you don't get a nice driver, you have to get off every hour or so and wait for another one. If you have to wait for a bus at three in the morning, you'll be waiting a long time. Anything can happen." Almost half the nation's major cities have laws against sleeping in cars and/or public places.

In Circle Six we traveled on the River Styx,
and passed the hordes of Hypocrites and Heretics
forever mired inside a sticky boiling mix
of silk to spin their tangled web of politics.

Said the governor: "When you send power back to the local level, the level closest to the people is generally best." He then went on to support a law that prevented local governments from approving paid sick leave for workers. Meanwhile, as the nation was spending increasing amounts on national defense, and as its financial industry celebrated a tripling of the stock market, and as just two men made more money in a year than the total cost of food programs for the poor, politicians voted to cut the food programs.

In Circle Seven was the gaseous flaring ground,
the fractured mountains, poisoned rivers -- a degree
of violation unforgettably profound,
the precious motherland exposed to Sodomy.

A remarkable human response to environmental damage has been observed: Those demonstrating the least sustainable behavior are least likely to feel guilty about their environmental choices. Many of the super-rich, well aware of the devastation around them, are taking no chances, isolating themselves with underground survival bunkers with 9-foot walls and bullet-resistant doors and sniper posts and trip sensors and pepper spray detonators and geothermal power units and anti-chemical air filters and infrared surveillance devices.

In Circle Eight the shrieking Furies dared to tread
in pits that quaked with voices of the living dead,
whose flesh was offered to the lash at tempest speed
and ripped to shreds before the Fates could intercede.

The nation's 'leveragers' use borrowed money to buy a troubled company, and then they pay themselves hefty fees, load debt on the purchased company, cut jobs and wages and pensions, cease maintenance on manufacturing facilities, and demand government tax breaks to keep the purchased company running. Other leveragers use government subsidies to buy up foreclosed houses, hold them till prices appreciate, and in the interim rent them back at exorbitant prices, sometimes to the very families that lost their homes.

We passed through Pluto's vaults to enter Circle Nine,

a sudden change to frozen depths containing souls

of those who profit from despair; to them a shrine

was fashioned, and their bodies wedged in icy holes,

and just beside them, fittingly, was Lucifer

himself, interred inside the icy sepulcher

with financier and warrior and governor,

their hellbound triple countenance to long endure.

In the past four decades, the nation's food costs have gone up by 100%, housing 250%, health care 500%, and college tuition 1,000%. Wages have stagnated over the same time period. Many people are suffering "deaths of despair" -- death by ***drugs, alcohol and suicide***. Drug companies have been eagerly promoting pain medications, and tens of thousands of people have died from overdoses.

And in the end we traveled up the mountain, far
from glacial blazing hell, from Lucifer's domain,
from scenes and spectacles perplexing and bizarre
and Plutocratic and satiric and profane.

Quest IV -- The Children: First Game

The rackety chattering of birds startled me from my slumber. A resplendent white sky, perhaps of mid-morning, forced my half-open eyes to the soothing darkness of the ground, and the unusually chilly air coaxed my arms around my trembling midsection.

Alan. He was gone. The realization wrenched me to my feet, and immediately I was calling his name, stepping in and out of a nearby pine grove, scanning the grassy landscape in front of me. I was some distance from the garden, within sight of the newer Logic Institute buildings. I had apparently been lying on the soft earth for most of the night, warmed by the bulky sweater acquired from our robotic guide. An empty water bottle lay to the side.

Alan, I conjectured, must have awakened early and decided to approach the Logic Institute entrance. Odd that he hadn't waited for me. My repeated calls met only silence, except for the derisive cawing of a nearby blackbird. At the end of the stately hedge walls from which we had negotiated our departure, about 300 meters away, glistening white in the morning sun, was the entrance to the building that we had so doggedly pursued. But still no sign of Alan.

I waited for quite awhile. Apprehension was building inside me: what could have happened to him? Had we been separated after our automaton-guided transfer from the garden? Memories of our convoluted passage from the tangle of greenery were already blurring in my mind. Thick, fast-moving clouds and a faint whistling in the wind now added a note of eeriness to the openness around me, and as a light rain began to fall I decided to walk to the school buildings to inquire about Alan.

As I passed along the outer hedge wall and approached the school, thoughts raced through my head: How did I get from the garden to the pine grove? I wish I hadn't left my phone in my backpack. It probably wouldn't work anyway. Alan didn't have a phone either. Where were all the Logic Institute students?

At the front steps of what appeared to be the main building, I finally encountered another person. A young boy, not more than eight or ten years of age, in a uniform of sorts, but displaying the disheveled manner of a child at play, appeared at the white double doors with a mischievous look on his face, as if deciding whether to risk punishment by leaving the premises.

"Son," I called to him, apparently catching him by surprise, for he whirled in astonishment upon hearing my voice. "What school is this?"

The boy evidently sensed my unthreatening tone, for he responded without hesitation. "I'm not sure. We just came for the summer classes."

"Do you know who's in charge here?"

"Miss Lovelace. She's the big lady at the end of the hall." He pointed inside, down a dully-lighted hallway that may have been on some sort of summer energy schedule.

"Thank you." I peered inside, as he gingerly stepped past me, but a sudden impulse prompted me to turn to him again. "Son, have you seen someone who looks like me? Dirty clothes and everything?"

"I saw someone earlier."

"Where? Can you help me?"

He hesitated, began to speak, then abruptly slipped back inside the building with a slight gesture for me to follow. We walked together down the spacious and ill-lighted hallway, the little boy quickening his diminutive steps in order to rid himself of this unexpected responsibility, and I tingling with expectation as the moments pounded by.

"I think he might be in there." He pointed now at a classroom -- or so it appeared -- near the middle of the corridor.

"Son, do you...?" But he was no longer at my side. Apparently changing his mind about his intended defiance of the rules, he bounded up a stairway to the safety of some nearby sanctuary. I cautiously opened the classroom door and peeked in, and to my surprise encountered a group of children, of early elementary school age or thereabouts, conferring in the corner of a small playroom with that simple earnestness peculiar to very young people. Immediately upon my entrance, before being allowed the opportunity to inspect my surroundings, the youngsters shrieked in delight and jumped up to greet me, all the while flailing their tiny arms and legs like so many wag-tailed little dogs.

"Mister, will you play 'do the loop' with us?" A doll-sized little girl with perfect amber skin appeared to be the spokeschild for the group, and the only one capable of controlling her giggles for the better part of the invitation.

"Children, I'm sorry, I..."

"Please, mister. Inky isn't here. We need one more."

The other children, encouraged by their intrepid little leader, bounced and cheered and tugged at my arms while pleading with me to join them. Perhaps because of the unexpected frivolity and my desire to appear agreeable, I allowed the nine or ten wee pairs of hands to usher me into a vast gymnasium area labeled 'Von Neumann Center,' with tile floors marked for indoor amusement and cathedral-high windows wire-meshed against the assaults of their young antagonists. Along the entire length of the other three walls inexplicably appeared a series of narrow openings, each about two feet wide and four feet high, perhaps two hundred and fifty in all, which may have better served some unorthodox spirit of aesthetics than of functionality. But in any event these openings provided a dimension of mystery and camouflage for the children, who, I imagined, would soon entice me with a round of 'hide and seek' within the slotted walls.

"I'm Addie." The little girl had stationed herself in the center of the gymnasium floor with the others around her. "Here's how you play. Lodie and me and you and Biffy are the commanders. Lodie fills the basket with the balls in the first room and then one of the other kids brings me the basket to dump in the big box. You take Inky's place and push Lodie to the second room so he can take an empty basket and do it again. Biffy turns the lights from green to red as fast as she can and tells us to stop when all the rooms are done and all the lights are red."

"But," I objected laughingly, while finding myself suddenly engrossed in the game, "there's a couple hundred rooms -- how can you count them all with only eight lights?" I spoke of a row of eight double lights, of the size and appearance of traffic signals, attached to the wall behind me and controlled by a set of on-off switches.

"No, no," Addie responded impatiently. "You don't count like that." She proceeded to explain a coding system whereby the first light, when switched from green to red, represented a 'one,' and the second light represented a 'two,' but a three was represented by the 'one' light plus the 'two' light. The next light represented a 'four,' so that a 'five' would be the 'one' light plus the 'four' light, and a 'six' would be the 'two' light plus the 'four' light, and a 'seven' would be the 'one' light plus the 'two' light plus the 'four' light. The light after that represented an 'eight,' so that any number from zero to fifteen could be expressed by some combination of the 'one-two-four-eight' lights; and the light after that represented a 'sixteen,' so that any number from zero to thirty-one could be expressed by some combination of the 'one-two-four-eight-sixteen' lights; and so on. In this way, of course, many more than eight rooms could be counted on the set of lights, although my efforts to determine the remaining sequence were thwarted by the childlike screeches that signaled the start of the game.

"Wait, children," I pleaded, but it was too late. The little boy named Lodie had already positioned himself on a small four-wheel cart at the first opening in a nearby corner of the room, and in another moment had surrendered a basketful of balls to another waiting child, who in turn rushed them to the center of the gymnasium floor, where Addie deposited them in a large storage bin as all the lights turned green to the sound of a metallic clanging signal, while yet another child started the short run back to Lodie with the empty basket. I stood immobilized during this commotion, fully expecting Lodie to roll himself towards the second opening to continue with the next waiting basket, but instead the procedure stalled as the children simultaneously ceased their activities to stare indignantly in my direction.

"Inky goes next," said Addie. "That's you now."

I couldn't fully understand the purpose of my expected role in the game, for Lodie sat only three feet from his second entry point, but I discarded my attempts to understand the rules in the face of the undersized but unwavering frowns focused at me from various parts of the room. Discretion seemed to dictate my joining the game if any information were to be subsequently gained from the youngsters. I breathed deeply in resignation, then sprinted to the nearby wall to the waiting Lodie, much to the uproarious delight of the revitalized children. Once there I grasped Lodie's cart by the ends and in one continuous motion pushed him into position at the front of the second wall opening, where, to the accompaniment of heightening shrieks of encouragement from his cherubic playmates he plucked the basket from a waiting hand and pulled himself through the miniature doorway, reappearing a moment later with a new basketful of balls for another waiting child, who rushed them to the center of the floor as the bell clanged and one of the lights turned red and I began to push Lodie to the next waiting basket at the third entrance. Within seconds we were proceeding at a furious pace: Lodie filled a basket; a runner whisked it to Addie, and she counted the balls before dropping them into the bin; another runner raced back toward Lodie as I delivered him to the next doorway. Everyone worked in swift harmony, including Biffy, who meticulously adjusted the lights to the flow of the ball movement and sounded a gong at the successful completion of each round of activity. Faster, faster we advanced around the exterior of the vast playroom: load a basket, add the balls to the storage area, increase the room number by one, listen for Biffy's signal; children racing to and fro, baskets back and forth, now here, now there; repetitious clanging noises marking our staccato movements with increasing frequency. I paused once, for just an instant as Lodie plunged through another doorway, to glance at Biffy's lights, which glared from the far wall in bewildering patterns: first 'green-green-red-red-red-red-red-red,' then 'green-red-green-green-green-green-green-green.' The process was continuing at a breathless pace, so that after another few seconds any hesitation would have risked the total disruption of the game -- if our bizarre activity could reasonably be called a 'game' -- or transformed the clocklike simultaneity of movement into the bedlam of an orchestra without direction. I concentrated on my personal, rather tedious role in the procedure, that of transporting Lodie from door to door, and after a while began to enjoy the soothing, monotonous rhythm of the repeating steps. Push, pause; push, pause; push, pause: so regular, so pulselike was the pattern that our progression along the side wall of the gymnasium required little conscious effort.

On and on we went, until finally, as I pondered the recreational content of our methodical movements and happily anticipated the fast-approaching final opening, a dramatic clangorous racket suddenly froze me against the wall in momentary panic. The shocking sensation subsided just as quickly with the realization that eight red lights glowed authoritatively from the wall ahead of me, even though one last wall opening remained unexplored by Lodie. It seemed for a moment that the fiery glow of the lights had been transferred, somehow, to the eyes of the children as they all turned to look at me.

"It's the new man's fault," yelled one of the basket-carriers.

"Yeah, we should've waited for Inky," said another.

"He ruined our game."

"No, he didn't," retorted Addie. "It's OK. We'll try again tomorrow."

With that the dejected youngsters replaced their game materials in the corner of the room and, without a word to me, filed one by one through the exitway. I paused in doubt for a moment, then followed Addie back to the school's main hallway.

"Addie, I'm sorry if I caused a problem."

"That's OK, mister."

"Please tell me – where are all the college students?"

The tiny girl looked up at me and responded immediately. "All the higher grades are upstairs. But no one's supposed to go up there."

I wanted to inquire further, but she had already scampered through a double door into a dormitory-like corridor which seemed, in my better judgment, beyond the allowable level of intrusion. Instead, the nearby stairwell lured me to the second floor, where despite the barrenness of the long corridor confronting me I resolved to remain until by some unknown means the necessary information about Alan could be acquired. Discretion was demanded, but so too was the sense of urgency that was building inside me.

Quest V -- The Children: Second Game

In darkened corners of the Institute,
outside a clamorous gymnasium,
with children gathering for some pursuit,
I wait in silence so I won't become

the jester's sacrifice. A strident clang
accompanies a strange scenario
of interactions, with the yin and yang
of teamwork here and there, to come and go,

and then, as smallish outlines multiply
nearby, amidst a colorful array
of naughty words - including 'crucify' -
I find an exit door and slip away.

The second floor appeared deserted, and the lighting subdued, as if indeed some weekday restrictions applied to the area. Silence surrounded me as I timidly entered the main corridor from the stairwell and weighed the feasibility of creating a clamor to attract attention. Better, I thought, that I approach doors at random. No other alternative seemed to exist. My harried thoughts returned to the peculiar game just played in the gymnasium, which seemed remarkably worklike to me, although the youngsters participated with all the zeal of kittens battling for a ball of string. And, too, they had treated it with such seriousness. Yet I had little time for analysis, because within seconds of my departing the stairwell there developed a drumlike, rumbling noise ahead of me, as of innumerable hoofs pounding the floor in a rapid approach towards the end of the hall. I froze against the corridor wall, uncertain what to do. In the next instant the sound was revealed as that of an advancing swarm of young people which converged upon me so hurriedly and in such numbers that my awestruck response to its appearance lasted but seconds, too little time in which to consider its purpose. The intensity of the multitude, and the inpenetrable thickness of its ranks, forced me back downstairs to the first floor hallway, where without hesitation I slipped behind a projection in the wall in order to watch their approach. They proceeded, thankfully, away from me and towards the gymnasium entrance across the corridor. From the indistinct shadows of my concealed corner in the wall I marveled at the implausible dimensions of the stream of children that passed by me. Five or six abreast upon their descent from the stairway, they required several minutes to funnel through the gymnasium doorway, so that by the time the final stragglers popped from the stairwell to overtake their classmates, I had estimated their numbers at about five hundred and their ages at perhaps ten to twelve.

Although most certainly some of the children had detected my frantic retreat from the floor above, they seemed undisturbed by my presence, and wholly unaware of my interest in their activities. After a brief recuperative period alone in the hallway, I collected myself for the few short steps to the gymnasium, where, I reasoned, I might both satisfy my curiosity about the children and inquire about Alan. To the former end I stole noiselessly to the entranceway and peered inside. Apparently another game of sorts was being arranged. Around the exterior of the vast room, beside the numerous wall openings that had provided the access points for the ball-retrieval game of the earlier group, an unbroken circle of children lined the walls in such a way that a single child was stationed at each opening. Within this outermost circle, about ten feet further towards the center of the floor, stood another smaller circle of children. Within that circle, and again ten feet further in, stood yet another circle. Within that another, and then another, and another, until finally a squarish loop of four children surrounded a pair of children who stood at opposite sides of, and a few feet away from, a single little boy in the middle of the room. All the children carried sizable baskets except the boy in the center, who stood beside the huge storage bin from the earlier game. Before I could attempt to identify the nature of the proceedings -- a doubtful accomplishment in any event -- the familiar, tumultuous clanging noise suddenly threw the outer ring of children into a frenzy of activity in which the wall openings were invaded by the youngsters stationed just outside, who emerged a moment later with some indeterminate object or objects in their baskets. In the next few seconds the pattern of the game was revealed. Pairs of children in the large outer circle converged upon single children in the next inner circle to combine the contents of their baskets -- small balls, again, as it turned out -- into single baskets. With one frantic, synchronized burst of motion the number of baskets carrying balls had been cut in half as each child in the second circle had accepted, and apparently counted, the balls of two children in the first circle. Immediately following this, and with the same coordinated oneness of movement, each pair of children in the second circle advanced towards a single child in the next inner circle to combine their balls into a single basket. I wasn't certain, but it seemed each child emptying a basket was obligated to call out the number of balls surrendered, so that each recipient of two basketsful might perform a quick calculation to determine the new quantity of balls. At any rate, the process continued, smoothly and rapidly in wavelike bursts from one circle of children to the next, so that within a matter of seconds all the balls had been emptied into the baskets of the two children just outside the center of the floor, who strained and staggered under the weight of their loads to their lone companion at the storage bin. There they shouted some indistinct numbers and fell back to the gymnasium floor as the discordant signal bell clanged a final time. The whole process had required just a fraction of the time taken by the earlier game.

I hadn't noticed a young boy standing next to me during the frenzied seconds of action inside the gym, but now as the game's end shifted my attention back to my immediate surroundings, I literally spun around in surprise when our eyes made contact.

"Hello, sir." He seemed courteous beyond his ten or eleven years, although understandably shy, and an image of freckle-faced innocence. He stood watching me, sipping slowly from a bottle of water.

"Hello, son. Why aren't you playing?"

"I got here too late. If they play again I will."

"What kind of game is that?"

The boy squinted at me in the obscure light of the doorway, as if attempting to better identify me. "You don't know? I thought you were one of the teachers."

"No," I responded awkwardly. "I'm a college student."

"That's parallel."

"Pardon me?"

"That's the name of the game. The little kids play 'sequential.' They call it 'do the loop.' In our level they teach 'parallel.' Our class can do it faster than anyone else."

"That's very good." I sensed a possible source of information in the child. "Is it fun?"

"It doesn't have to be fun. We just have to be good."

"I see. Tell me, where are all the teachers?"

"Our teacher is Mr. Blaise. He had to go to the city."

I struggled to assemble a functional sequence of words for my next question. "I'm looking for another college student who looks a little like me. Do you know where all the adults are?"

"Maybe he's in the logic class."

"The logic class? Where is that?"

"I don't know. In the other building, I think."

Disappointment filled me again as it became apparent that little had been gained by our conversation. The boy, perhaps in an attempt to escape the pressure of responding to my inquiries, quietly opened the door and slipped through to join his classmates, leaving me the lone observer in the hallway. I decided after a few moments' deliberation that my presence at the gymnasium served no useful purpose, for apparently the children would remain occupied for some time. Furthermore, it seemed likely that the simple confirmation of Alan's existence would represent the limits of their assistance. As a result, I turned and hurried up the stairway to the second floor, where my next few minutes were devoted to a methodical search of a dimly-lighted corridor that seemed much more extensive than the main hallway, as if it somehow continued into another building. A number of partitions divided the lengthy corridor into secondary spaces that were brightened by the eerie glow of yellowish, overwaxed floors. I entered each area in turn, knocking on a few doors to no avail, and soon realized that the lengthy silent hollow space in front of me promised little of interest. Yet the notion arose within me that some individual would suddenly appear to guide me in serendipitous fashion to Alan.

Quest VI -- A Strange Man

As if proceeding down a rabbit hole
I come across an enigmatic soul
who tries in vain to make me understand
the patterns of a neural wonderland.

An indescribable sensation swept through me as I tried to
reconstruct the rectangular patterns that had led me away from and
then back to the main corridor. I felt that I had traveled such a path
once before. As unfamiliar as the building was to me, some far-off
sense of intimacy preyed upon me as I wandered through the
empty halls, as if a childhood memory were being stirred from its

cobwebbed niche deep inside me.

Yet finally it became apparent that I had simply been retracing my steps through the directionless hallways, and that my search should continue elsewhere. Somewhat dejectedly I returned to the head of the main stairwell with the intention of checking the progress of the children in the gymnasium. Before I descended, however, my attention was diverted by a thin crack of light from the base of a nearby doorway. I carefully approached the unlabelled entranceway. Not a sound was evident, not even the faint rumblings of the counting game from the gymnasium below. Impulsively, before allowing myself time to reconsider, I entered the room. Before me appeared the characteristic furnishings of a faculty lounge: tan leather couches; numerous lamps and stands of remarkable square white plainness; handsome cedarwood bookcases. Only when I ventured further inside was I able to view the nearest corner of the room, where, rather startlingly, a little bespectacled man stood to address me.

"Hello, my good man." I stumbled back against the door upon hearing his voice. "The name's Dootil. D-O-O-T-I-L. Join me, won't you?"

"I...sure, I'd be happy to." I stepped toward him, eagerly accepting his offer of a chilled bottle of water.

"Another lamb to be shepherded, eh? You seem a little older than the rest of them. Ah, the basic income allows all ages to pursue their dreams. What do you plan to study, my boy?"

"I, uh..."

"Ah, to share the fruits of knowledge with our youth. I myself instill their growing minds with the facts about the machinery of the brain. Have you read any of my work?"

"Uh, not yet. Listen, I'm looking for..."

"Yes, I'm content here, guiding the young sculptors of our future world. Enlightening them about the chemicals and neural processes that make up their minds. Thoughts reduced to neurons. It's quite an astonishing hypothesis, isn't it?"

I couldn't combat his rapid-fire manner of discourse, so I waited patiently to ask about Alan. The man was slight of build, but filled with energy as he spoke, like an overprotective ground squirrel chirping away at a sudden trespasser. His balding head turned a deep red in the very midst of his outburst, so that I almost expected to see steam escaping from his forehead as he went on.

"There's no magic to it. All this talk about a ghost in the machine. Hah! Here, sit down, I was just rehearsing my autumn lectures."

He waited, as if expecting me to acquiesce to his command. But I was looking for a tactful way to excuse myself. Still, a brief rest would do me good. Perhaps the man would listen to me in a minute or two..

"Now how does the brain capture the image of a red square and a blue circle without mixing up the colors and shapes? First of all, 'red' and 'blue' and 'square' and 'circle' are each represented by a neural network. Another neural network binds 'red' with 'square' and another one binds 'blue' with 'circle.' That's spatial binding."

"That's interesting."

"You can 'picture' the whole image of a red square and a blue circle together, but the mind can consciously focus on only one colored shape at a time. When you try to picture the two colored shapes side by side, that imagined scene is unfocused, vague, a simple image lacking in detail. Try it yourself, my friend. It's easy to envision a red square all by itself; it's easy to envision a blue circle all by itself. But picturing the two objects together just results in a type of 'photograph' in which all the objects blend into a single display, with no one object standing out. Any claim to consciously focusing on both objects at the same time is an illusion, caused by the mind rapidly switching from one object to the other."

Dootil paused, looking quite proud of himself for his eloquent oration.

"Now," he continued, "a listener might ask, if one cannot truly imagine two objects simultaneously, then how can 'red' and 'square' together be imagined so vividly? Because it's a composite unit. Two neural networks bound by a connecting neural network. Same for 'blue circle.' On the other hand, the imagined picture of a red square and blue circle *together* is simply a photo-like image, a neural network without composite units, an approximate reconstruction of a previously perceived or imagined scene. It is not a consciously focusable unit."

He paused again, and I should have taken the opportunity to interrupt him, but I found myself strangely captivated by his impassioned rendering of his brain hypothesis. I waited as he rustled through his paperwork to produce an image that seemed to portray interconnected squares of neural matter.

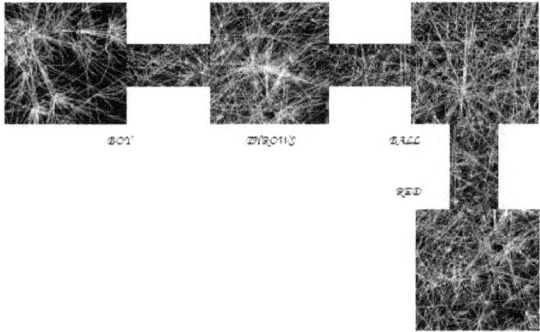

"Now for something more complex -- 'The boy throws a red ball.' That's two composite units, 'boy throw ball' and 'ball red.' Of course the whole scene can be pictured, but, as before, in a manner that does not focus directly on either of the composite units. Any attempt to focus simultaneously on the act of throwing the ball and the physical qualities of the ball is either a concurrent process, with the mind switching back and forth in millisecond time between 'throwing' and 'redness,' or a return to the complete image, in which the mind is not consciously directed toward either 'throwing' or 'redness.'

"With two composite units, how does the brain consciously focus on one or the other? Through temporal binding, of course. For 'The boy throws the red ball,' the neurons encoding the composite unit 'boy throws ball' are firing at a standard activation rate -- perhaps 40 cycles per second -- while the neurons encoding the composite unit 'ball red' momentarily cease firing, or perhaps continue to fire at a slower rate while remaining synchronized -- perhaps every other pulse -- with the faster neural network. For 'The ball the boy throws is red,' the activations are reversed, thus giving precedence to the composite unit 'ball red.'"

Now he paused for several seconds more, breathing heavily, and I began to rise to demand some help just as he blurted out a new thought:

"The problem is, the boy may be a girl and the ball may be blue. You can change what you see to suit your own reality."

"Your own reality?"

"Yes, we can have our own realities. Reality and imagination exist in the same part of the brain."

Dootil seemed pleased that I appeared interested in his discourse. "Let me show you where we do our work," he promptly said to me. "It's the most advanced and automated lab you'll ever see." He continued, as we left the lounge and climbed a nearby stair, "there is much of curiosity for me to show you there."

"No, please. I'd very much like to, but I really can't. You see, my friend is waiting for me."

He continued up the stairs and through a third floor entranceway as if unaware of my comment, or unconcerned with it. As we turned onto the main corridor I noticed a hallway door across from the stairwell marked 'TO LOGIC STUDIES.' Immediately, and perhaps too abruptly, although Dootil seemed sufficiently inattentive to my actions, I stepped through the doorway and bounded down a set of stairs into another passageway that presumably led to a more advanced training sector of the school. As before, the hallway lighting had been reduced to a minimal level. I looked ahead to attempt an identification of the area ahead of me, but the dullish light blurred my view and created a disturbing illusion of emptiness, such as one might experience on an unfamiliar road before dawn.

I was relieved to rid myself of Dootil, whose strange mannerisms induced shudders of repugnance inside me long after his departure. But my mood had worsened nonetheless. The eerie atmosphere of the school had so unnerved me that a momentary glimpse of shadowy light beside me or a faint creaking in the walls was nearly intolerable. I became aware of every intrusion upon my senses, no matter how slight -- and many that were fashioned in my own mind. Memories of boyhood adventures in abandoned houses unavoidably returned to me: ageless mounds of debris welded together with rusty nails and bits of broken glass; the pungent odor of damp, decaying wood from weakened floorboards. Most of all the anticipation of ghostly horrors that seemed so certain, but thankfully weren't to be.

I sat on the floor to rest. I had never felt so lightheaded, and tired. In the muted darkness of the hallway my thoughts gravitated to Dr. Babbage's language research. Dootil's ramblings may have stuck in my mind. An odd time for neurophilosophy, it seemed. But I had read Babbage's treatise before we began our travels, and it seemed that a mental review of his work might serve to reignite my enthusiasm for the wonders of the Logic Institute. My racing mind tried to organize his thoughts on intelligent robots.

Robots seem potentially capable of any activity, any thought process, any simulation of humanness. But there are some fundamental differences between humans and machines. Among its billions of neurons the human brain can accommodate an unending assemblage of neural networks; these are dynamic structures, constantly changing in their relationships to other neural networks through synaptic formations that are strengthening and weakening amidst an infinite flow of cognitive experiences and sensory input; with neural charges that are multi-state rather than binary, their levels of activation varying with changes in competing or reinforcing neural networks and the passage of time; and with spatial properties that allow them to participate in any number of neural networks in related and unrelated mental processes. Any simulation of all this will likely have to evolve from a simple infant-like artificial neural apparatus. But any such experimental and expedited and random evolution may never result in a humanlike state.

I felt exhausted, both physically and mentally. And I was hungry. For a few moments -- or perhaps longer, for the dingy and unchanging hallway made it difficult to judge the passage of time -- I slept, as hallucinogenic-like outlines of red balls and robots and doorways and Dootil took shape in my mind. Then came a spontaneous flurry of multi-colored neurons firing in a complex pattern to recreate an image of the hallway around me, which remained as a type of mental photograph, imprinted in my mind as my brain formed new synaptic connections among neurons making up the image, and made more permanent the longer the image remained the object of my attention.

Quest VII -- Logic Class

As class convenes I quickly indicate in clear
logician's diction of my eagerness to share
my disposition to determine if we're here
or there, or if instead we're neither here nor there.

A student dissident, who would have been dismissed
by any academic dean or exorcist,
got up and shook his fist, proposing to assist
by posing logic twists too cryptic to resist:

"If you were here or there then if you weren't here
you would be there; if everybody isn't there
you'd certainly be here; but if you aren't here
and aren't there you can't be here or there - it's clear

you can't be anywhere - but if you aren't there
or aren't here you could of course be anywhere,
for then you could be here, which isn't there, or there,
which isn't here, unless you're neither here nor there."

And somewhere in the list of dictums missed amidst
this syllogistic mist there did indeed exist
a philosophic gist that no one could enlist
since common sense insisted class should be dismissed.

I awoke with a start, and after a moment of orientation forced myself to emerge from my secure little corner of the semi-darkness to continue my search for Alan. I hurried down a lengthy corridor with whitish wall tiles and a well-worn floor, then turned a corner and heard voices in a nearby classroom. As I looked inside the doorway, the differences in the ages and number of children between this and the other classes, and in their particular forms of recreation, became immediately obvious. These people were young adults, or nearly so. They were generally more subdued than the frolicsome sprites of the earlier sessions. Four of them stood near the entranceway, sipping from bottles of water as I stepped through the door.

"Pardon me." I spoke to no one in particular. "I'm looking for my friend." Many of the teenaged youngsters stepped back to allow my approach, while others stared at me with a mixture of curiosity and amusement.

"Who?" A studious-looking boy whose heavy glasses forced his ears away from his head responded with a peculiar grin.

"This is the logic class, isn't it? I was told my friend might be here."

"Can you describe him?"

My spirits sunk again, and a feeling of irritation gathered inside me, but I managed to maintain my composure. "Well, he's thin, and not very tall."

"You mean he's not tall and thin?"

"No, I said he's thin."

"I said that too."

Confusion silenced me as a chubby boy with a lisp interjected his opinion. "Then we can't say he's tall or not thin."

General agreement seemed to be reached on that point, for after a few seconds six or eight of his fellow students murmured in support of his apparent clarification. I noticed for the first time that a small crowd had assembled around me, the greater part of which consisted of spirited and untidy teen-aged boys, of whom the studious-looking one appeared to be the most outspoken.

"Look" -- I tried to impress a note of seriousness upon them -- "do you know someone like that?"

"It depends," said the studious boy. "If we know a thin, not tall person, is it definitely the one you're looking for?"

"Yes!"

"In that case," added another boy, "if it's not the one he's looking for, he can't be thin and not tall."

"But we might have more than one like that," said someone else.

"No, what he's saying," the studious boy announced, "is that either it's not the one he's looking for or he's thin and not tall."

The others stopped to think about it, as I shook my head in frustration.

"He's right," said the chubby boy with the lisp, and a few others nodded supportively. "If it's not the one he's looking for, he must be tall or not thin. But if it is the one he's looking for, he has to be thin and not tall."

"Damn it, where is he?" I immediately regretted my loss of temper, and forced a smile in an awkward attempt to make up for it.

"He's in the Logic Institute," responded the studious boy. "That's where you have to go."

"But I'm there now!"

"How can you be there if you're here?"

I closed my fists and tried to control my quivering chest and arms.

"You see," the boy continued calmly, "if you were here or there then if you weren't here you would be there, although if everybody isn't there you'd certainly be here. But if you aren't there or aren't here you could be anywhere -- for then you could be here, which isn't there, or there, which isn't here."

I barely heard his final words, or the laughter of his friends, for I had already turned away in search of more lucid representatives of the logic school. As I followed the main hallway I encountered a narrow corridor whose distant end revealed another doorway, and whose left wall was intersected, about half-way down, by a doorlike opening. From this latter point emerged the boyish, shrill-sounding echo of a solitary speaker and the murmurings of an appreciative crowd which occasionally burst into a frightfully riotous applause. Cautiously I approached the side entrance and peered in. A dark-haired, intense-looking youngster of about fifteen stood before a blackboard marked 'WORK GROUP 16' in heavy chalk letters, and behind a wooden lectern which bounced and swayed with each animated gesture of his body as he delivered his speech, as if the floor were convulsing beneath his feet. An audience of perhaps fifty young people of approximately his age -- certainly a lesser number than I had anticipated from their boisterous reaction of moments before -- sat attentively as their discussion leader worked himself into a mild frenzy of speechmaking.

"Truth will win out!" he exclaimed, with a resounding note of self-confidence. "Try to trust in this thought -- I insist upon it! You must work with this philosophy, or your growth will stop without your knowing it. Only through truth will you triumph! So work vigorously, illustrious group -- morning or night, truth is still your only tool. You must shout this worthy wish on high:

'In living with truth,
I'm living with honor!'

So now your poor, your strong, your right, your wrong, will join in song:

'I worship you, oh glorious truth!'"

Despite my ignorance of his meaning and the purpose of the work group, the intensity of the youthful speaker's words drew me from the shelter of the corridor wall, so that to my dismay he noticed me as he was acknowledging the crowd's enthusiastic response to his speech.

"Why, who is this? Join us, sir!" The others quickly looked around to identify the object of his attention, and I reluctantly stepped from the shadows of the back wall to greet this new congregation of students. Apprehension had built inside me, but nevertheless I was warmed by the speaker's initial gesture of friendliness.

"My friend is in the graduate school. Do you know where I might find him?"

"Know who?" One of the students returned the words as the greater number of them rose to gather around me.

"His name is Alan. A graduate student. Dressed like me. Have you seen him?"

"Not in this room," responded one of the youngsters.

"Sorry, I know nothing," said another. They seemed confused by my question, as if it shouldn't have been asked. I studied their perplexed faces and felt the tingling of expectancy in my chest fade to disappointment.

"This is not right," exclaimed the boy who had delivered the stirring lecture. "His using 'A' in our room."

"Or 'E,'" said another.

"You must stop this, sir," the boy said to me; "I insist upon it!" And the others mumbled in agreement.

"What did I do? What do you mean?"

"Oh monstrous! Stop him!" cried a suddenly distraught young girl, whose hands pressed to her ears as if to shield them from my venomous attack.

"Look, sir," the boy continued, as those around him either turned away in embarrassment or glared disapprovingly at me, 'A' will just not work; nor will 'B' or 'C' or 'D' or 'E' or 'F.' 'A' is to sum or multiply; so is 'B' through 'F'; so is '0' through '9.' 'G' is OK; so is 'H' or 'I' or 'P' or 'S' or 'U' or 'Z.' Not 'A,' though; not 'A' or 'B' or 'C' or 'D' or 'E' or 'F.'

"I don't understand..."

"Oh, horrors!" someone shouted.

"Throw him out!" pleaded another young voice from the rear.

"You must go now," the speechmaker directed me. "This will just not suit us." He literally pushed me back through the entranceway, and slammed the door shut behind me.

I stood bewildered, and thoroughly dejected, in the stillness of the dim-lighted corridor. Through the sealed doorway the muffled sounds of children turning to prayer accosted my bedeviled senses:

"Pity his poor unholy spirit..." Yet immediately I became aware of another noise behind me, as of the titterings and shushings of youngsters in a schoolroom. I whirled and fell back against the opposite wall, almost stumbling over two or three small children in the process. I immediately recognized the tiny group leader from the first game.

"Addie!"

Amidst the nervous giggles of children defying some late-afternoon regulation, she addressed me with a much-welcome sincerity.

"Are you still looking for your friend, mister?"

"Yes!"

"The knowledge machines can tell you where he is."

"The knowledge machines? Where are they?"

"On the third floor. But don't use the ones at the end."

"Can you show me...?"

"We can't go up there." She paused a moment more, and then mumbled an indistinct word of advice -- "They'll be cursed," or something to that effect -- before joining her mirthful friends as they clamored back down the stairway at the end of the main hallway.

I hesitated myself, wondering about the prospect of following the children, but then hurried to the other end of the hallway, which was deserted -- although the motionless shadows cast in the area of the security lights added to my growing trepidation. With a burst of unanticipated energy I rushed up the stairway, reflecting at the same time on the levelheadedness of one little girl amidst all the lunacy.

Quest VIII -- Recursed

Door B, Door B, Door B, Door B, I hurry through
until those very doors begin ejecting me,
and then, as if invited into deja vu,
by some unlikely act of serendipity,

I hurry through Door A, Door A, Door A, Door A,
through blurry thresholds like a prison runaway,
while knowing not, amidst the feverish affair,
the purpose or condition of my being there.

The dreary surroundings at the head of the stairs on the third level called to mind the strange Mr. Dootil, and the roundabout escape route that had guided me to the indecipherable logic class. As I ventured further along the hallway and strained to see in the darkness, I noticed that indeed a row of apparently interactive devices lined the walls. Stepping toward one of them, I verbally requested a communications session, but received only a cryptic exhortation on a glowing screen:

WHATEVER OCCURS RECURS

I wondered if the knowledge system were temporarily disabled, for there seemed to be no reason for this machine to refuse to talk to me. Perhaps other knowledge devices would lie ahead. Without further deliberation I stole past a series of empty classrooms, straining again to distinguish the squarish features of a communications mechanism, and then abruptly encountering a doorway labeled 'Knowledge Room' near the end of the hall. I opened the door and entered a little area not unlike a doctor's waiting room, but unfurnished except for a few chairs lining the windowless white walls. At the far end another doorway, adorned only by the notation 'Room B' and the greenish glow of a nearby knowledge device, loomed as perhaps the last obstacle between me and Alan.

After ensuring that the Knowledge Room door opened from both sides, I hurried to the opposite end of the room, where the smooth molded glass of a knowledge apparatus faced me from a bare wall like the eye of a great motionless being.

"Hello," I half-whispered into the surrounding darkness.

"HELLO." Its response startled me, despite my instant delight at having initiated contact with the knowledge system. I studied the eerie green glow of the tiny letters that had accompanied the voice, and hurriedly organized my thoughts into a series of inquiring demands.

"I am looking for my friend." I went on to give Alan's full name.

An uneasy hesitation followed my words. When the machine finally spoke, it droned emotionlessly, with a heavy masculine undertone.

ALAN IS NOT AVAILABLE.

"How do I find him?"

YOU ARE NOT AUTHORIZED TO FIND HIM.

My lack of progress was, as usual, frustrating, especially in such a bleak corner of this outlandish place. I tried again, carefully articulating the words while adding some new information, and the machine finally responded with the same note of detachment in its unvarying tone:

YOU HAVE BEEN IDENTIFIED, PETER.
PLEASE PROCEED TO ROOM B.

I was elated by this unexpected sign of progress, but still filled with skepticism. The door to Room B opened up to another sparsely furnished enclosure of pale, windowless walls and a pulsating communications device. Receiving immediate instruction from the assisting contrivance, in the polite formalities of electronic voice transmission, to have a seat, I settled uncomfortably into one of the metal chairs that lined the barren walls. A jittery silence embraced me as I waited for further word about Alan.

Minutes passed without a response. Expectancy turned to

restlessness as the silence persisted. I felt increasingly warm. My sweat-dampened clothing was beginning to stick to my skin. I rose, finally, to steady my nerves and to relieve a compulsion to open the next door.

My breath halted momentarily when I noticed the 'B' on the door in front of me. The same notation had marked the previous door, leading into the room in which I now waited. Impulsively I spun around to look back. An 'A,' sensibly enough, marked the return to the preceding room. But the 'B' before me seemed to have been inscribed in error, for I unquestionably stood in the room earlier referred to as 'Room B.' The next room should be 'Room C.' I realized, suddenly, with an unintended smile of understanding, that someone must have mislabeled the door. The thought of the everpresent human capacity for miscalculation, even in the Logic Institute's supposedly precision-minded environment, helped to relax me.

As my returning glance passed the knowledge interface, I noticed a change on the screen: the message appeared longer. Immediately I lurched back to attentiveness, involuntarily releasing a grunt of acknowledgement while leaping towards the beaming communications device. A new message had indeed appeared:

CHECK PASSWORD.
ANNOUNCE VISITOR'S PASSWORD
AND WAIT FOR RESPONSE.
IF PASSWORD NOT ACCEPTED
PROCEED TO ROOM B AND CHECK PASSWORD.
RETURN THROUGH DOOR A.

"Visitor's password!" My trembling, high-pitched response rose unexpectedly from inside me. Instinctively I blurted 'Babbage' towards the knowledge screen, and a voiceless response appeared immediately:

PROCEED TO ROOM B AND CHECK PASSWORD.

The stifling, uncirculated air pressed against me and speckled my skin with drops of perspiration. I felt my chest expand and contract in rhythmic motions. The message made no sense to me. In addition to having had already entered a 'Room B,' I had perceived an absence of purpose in doing so.

I swung open the door marked 'ROOM B.' Inside appeared another waiting room, a virtual replica of the first two, with once again a 'B' on the door at the far end. Finally sensing some developing lunacy, I turned back in the direction of the outer room and the hallway.

But the door through which I had last entered was locked! With whiplike speed the realization hit me: my freedom had again been snatched away. I stood helpless and disoriented, taunted by the utter quiet of the tiny room.

My only recourse appeared to be cooperation with the accursed machine. I cautiously stepped into the next room, the third waiting area, where the inevitable knowledge interactor beckoned from the far end. I shuddered as I read its message:

CHECK PASSWORD.
ANNOUNCE VISITOR'S PASSWORD
AND WAIT FOR RESPONSE.
IF PASSWORD NOT ACCEPTED
PROCEED TO ROOM B AND CHECK PASSWORD.
RETURN THROUGH DOOR A.

Automatically my eyes rose to the door. The next room's 'B' label glared defiantly from its steel surface. I whirled around, knowing immediately what to expect on the door behind. The absurdity persisted -- each room replicated the one before, and each proceeded to another room 'B' while pointing back to a mislabeled room 'A.'

Without thinking I clamored into the next 'B' room, and into the same bare-walled white hollow that had enclosed me in each of the others. Again I encountered the shapeless metal chairs, the insanely constant door markings, the excruciatingly patient electronic servant bearing messages from an unknown master. Once more the enigmatic words confronted me:

CHECK PASSWORD.
ANNOUNCE VISITOR'S PASSWORD
AND WAIT FOR RESPONSE.
IF PASSWORD NOT ACCEPTED
PROCEED TO ROOM B AND CHECK PASSWORD.
RETURN THROUGH DOOR A.

I charged through the waiting 'B' door to the next communications screen. The inescapable 'password' command elicited a stream of curses from my lips as I hurried through another 'B' door into yet another waiting room, and then another, and then another.

Again and again I repeated the cycle. Again and again I passed through the doorways, each time pleading with my unnamed tormentor for mercy, until finally I succumbed to frustration and exhaustion. It had happened to me again. I had become a part of their demented game. Just like in the mountain pass and in the hedges. I collapsed into a hard-backed chair beneath the smirking gleam of the knowledge device, and as my sweat dropped in little trickles upon my protesting eyes, a surge of energy directed my foot up and against the machine's glassy face in a reckless attempt to extinguish its invisible spirit. I yelled for help, then fell to my knees and pounded on the wall until fatigue forced a return to calm.

After allowing myself a few seconds I tried to evaluate my position. The rooms were arranged linearly in a lengthy wing that probably extended north or east from the main building. I lifted myself from the floor in an effort to approach the knowledge screen, but the sudden dizzying movement made me stumble against it. As I lowered my head to regain my equilibrium, I felt the stifling heat of the room against my swollen eyes and throbbing temples -- a heat that seemed to have worsened as I progressed from room to room.

I made numerous attempts to supply the appropriate 'visitor's password,' but without success. All efforts to reason with the stubborn device met with silence. Finally I continued my passage through the corridor of doorways, thrice more enduring the meaningless "check password" message while encountering a ghostly sameness beyond every 'B'-labeled entrance. Unquestionably the temperature was rising. I paused to regain my breath in an ovenlike room that pressured me to quit while defying me to proceed to the source of the worsening heat. Each room inflicted greater hardships on my weakening body, yet each room urged me closer to some unimagined climax.

After long moments of uncertainty, I resolved to continue until I had escaped from the chamber of rooms. As I entered the next room, I pushed a chair against the opened door in a largely purposeless attempt to retain access to the room behind, but as a result the next 'B' door remained locked. Disgustedly I kicked the chair away, then continued into the next room, and then the next, where finally I was forced against the walls to avoid the intense heat that rose from the floor and simmered the moisture in my pores. Yet still I felt the need to press on. With my head turned to the cooler concrete walls, I groped my way to the next exit and flung open the door, and then pulled back with a shout of disbelief: surging beneath me, far below a cagelike arrangement of iron bars that secured the last doorway, lay a rugged, blusterous river that glistened with rocks and whitecaps in the shadowy afternoon light. Far below me it roared through the twisted valley, like a great replenishing artery to the devouring earth. Frigid breezes rose with the spray from its sputtering surface and clashed with the seething air behind me, leaving me interned between a steamy inferno and the icy violence below. I clung desperately to the iron bars that cooled and restrained me, the gateway to and from the Hades-like extremes, the fragile barrier between a prolonged and sudden end.

Abruptly a voice called out from behind me, and I turned to the communications screen as it flashed a command from its searing corner of the room. I focused sweat-stained eyes upon the blurred message:

PASSWORD ACCEPTED.
RETURN THROUGH DOOR A.

Confusion held me for an instant, for I had never uttered a password...but then I jumped to action. I pushed breathlessly on the door at the other end of the room. It opened immediately.

The heat slapped at my face, and the pressing heaviness against my lungs became nearly unbearable, but I battled another few steps to reach the next 'A' door, which yielded instantly to my touch. Excitedly I scrambled towards the next door, and the next, each time shielding my face against the scorching blasts from the floor and plunging through the opening like a routed barn animal. I could sense, even in my continued discomfort, a slight reprieve from the heat in successive rooms. I advanced with an energy that flowed more from my rejuvenated spirit than from my body: door 'A' to door 'A' to door 'A,' one after the other, faster and faster, until I finally outran the heat and the messages and stood once again in the musty hallway that some immeasurable time before had betrayed me to the torturous passageway.

I dropped exhaustedly to the remotest corner of the cool, lightless hallway's end, where I waited in silence in defense against any further assault. A number of seconds passed before I recognized the sound of a voice coming from the 'Knowledge Room' where I had encountered the first of the relentless stream of knowledge machines minutes before. Pressing my ear against the door, I heard a parting instruction repeated over and over again in a familiar, depressing monotone:

ALAN IS IN THE LOGIC INSTITUTE INFIRMARY.
GO TO BUILDING 2, ROOM 16 FOR ASSISTANCE.

ALAN IS IN THE LOGIC INSTITUTE INFIRMARY.
GO TO BUILDING 2, ROOM 16 FOR ASSISTANCE.

ALAN IS IN THE LOGIC INSTITUTE INFIRMARY.
GO TO BUILDING 2, ROOM 16 FOR ASSISTANCE.

I slumped against the nearest wall, stunned by the sudden revelation. Something had happened to Alan. Emotions burned in my eyes as I struggled for a more favorable interpretation of the message, but none was forthcoming. Alan had been injured in some way. I felt my head pounding as the realization sunk in.

Minutes passed in the secluded corner of the hallway before I regained control of myself. The faraway sound of a slamming door alerted me to the presence of others around me. Without a moment's consideration to the location of the noise, I started toward the other end of the hallway, and proceeded with such hurriedness that the evenly-spaced security lights seemed to flicker in rhythmic accompaniment to my pace. At the head of the stairs I paused, but heard nothing. Then, with the stealth of an escaping prisoner, I glided down the two flights to the main corridor, where my over-reactive eyes and ears turned one way and another to guard against the perils configured in my mind. I passed the gymnasium without daring to look inside, although the nonsensical strains of a children's verse overtook me from behind:

'Twas brilliant of the slithery toad
The fire to kindle beneath the waves.

I felt intensely emotional for a final moment before clamoring through the main entranceway and finding refuge behind a great oak tree, which stood in a nearly impermeable layer of leafy vegetation about fifty meters from the side of the building. The school's walls towered over me as I lay recovering my breath, their silvery whiteness brandished even more boldly now in the late afternoon sun. The brightness forced me to turn away, and caused patterns of light to race this way and that, meteor-like, against the black background behind my tightly closed eyes. After a while I began to feel soothed by the continuing silence and the warmth of the receding sun against my face. But I couldn't rest, for the image of Alan was still with me. I had to find Building 2, Room 16.

In the weighty stillness of the humid air I trudged through the tall grass along the driveway leading from the main building, following roadsigns to a smaller structure labeled 'Building 2.' I paused for a moment -- fears of the unexpected still lingered inside me. And I was skeptical about the nature of any assistance I might receive in Room 16. But the need to find my friend drove me on. Cautiously I approached the main entrance and stepped inside. The empty hallway had the brightness and contrived cheerfulness of a medical setting, and as I followed the numbered doorways it became clear that a group of medical admitting offices indeed lay before me. With uncharacteristic boldness I entered the designated room, where I was cordially welcomed by a white-clothed attendant with a nametag identifying her as "Candide," who claimed, curiously enough, to have been expecting me. Disregarding my flustered questions about Alan the kindly woman attempted to relax me with a chocolate bar and a bottle of water and affirmations of my friend's well-being. She assured me that Alan and I were in the best of all possible places. All too quickly I was transported to a state of serenity under her care, and it became increasingly difficult to remain alert to the words and actions of my benefactor. I battled the urge to sleep, but the punishing physical and mental pressures of the day had taken their toll, and so I allowed myself to close my eyes for a little while.

Quest IX -- Biochemistry and Madness

The vulgar throb and throes of evil lash
at me from deep inside the man: a beast
obeying primal calls to wail and slash
till all my pleas for clemency have ceased.

The frigid air in the room finally awakened me. I didn't know where I was. Such radical swings of consciousness had never happened to me before. My feelings of fatigue were laced with occasional bursts of mental energy that promptly melted into inertia. I waited now, barely able to move. Memories of the school's hallway abruptly returned, and the thought of Alan's disappearance sent a sickening jolt to the pit of my stomach. The clock above me said six-thirty -- whether morning or evening I didn't know. The room was unfamiliar to me. I lay on a cot in a sparsely furnished hospital-like facility, bright and sterile and noiseless. Hunger and tension gnawed at me from inside. I felt uncomfortably dirty and disheveled, and a bitter taste clung to the roof of my mouth.

Thoughts of Alan quickly re-entered my mind. I had to find him. My attempts to rise caused a painful dizziness in my head, so that after a moment of disorientation I found myself sprawled on the floor. Before long I steadied myself, although an audible throbbing persisted in the area of my eyes. I took a minute or two to examine my meager surroundings. The words 'L.I. Examination Room' appeared on a medical gown that hung near the door. This confused me – had I been transferred to another part of the hospital while unconscious? But that didn't matter if Alan were here too. After carefully lifting myself to my feet, I staggered out the single doorway to a lengthy hallway that connected dozens of little medical enclosures apparently just like mine.

As I started toward the exit door at the end of the hallway, the rapid patter of footsteps suddenly rose from a stairway just beyond it. Before I could evaluate the possible consequences of a surprise encounter, my position next to the door had blocked the entrance of a bearded, well-dressed gentleman whose face, in the darkened recess of the stairway, appeared more wolflike than human.

"Where are you going?" he asked. His gravelly voice accentuated his coarse features, and I looked with concern at his demanding expression while groping for a response.

"I'm looking for my friend. Can you help me?"

The dark-featured man frowned at me, then signaled for me to follow him upstairs, to a floor which might have been expected to contain a similar arrangement of hospital rooms. But instead, as we emerged from the stairwell, we entered a chemical laboratory of some type, apparently free of activity for the moment, although its normal procedures were unmistakably defined in the sour, permeating stench of acid, the endless black countertops lined with beakers and amber bottles, and the gurgling hiss of a nearby suction apparatus. The man led me past a mesh of tubes and hoses to a single office at the far end of the lab. There I waited with considerable apprehension as my host arranged himself behind a handsome metal desk and directed a cold, penetrating stare toward me.

"You're looking for Alan."

"Yes -- how did you...?"

"He's in the lower level. He came here looking for you after you walked away from the garden. I guess you had a little too much to drink!"

He chuckled to himself, then turned to his knowledge device and mumbled a series of commands, presumably in regard to Alan's whereabouts. The man had an intensity about him that could have been mistaken for delirium. His impassioned demeanor, the telltale widening of his fiery eyes as he spoke, the bullish determination that glowed pink from behind his unruly white beard: all this contributed to a fearsome presence that dominated my attention as we sat together, even to the exclusion of more pressing concerns about Alan. I had noticed his name and affiliation on the door - Victor Golem, Director of Medical Evaluation Services. But the significance of his role remained far outside my ability or willingness to understand it. I wished only to acquire information about my friend's condition.

I had just directed my attention to a book on the man's desk that was labeled 'Job Log' when he abruptly swung the knowledge machine toward me. An uneasy silence followed as he simply stared at me without uttering another word. His face was set in a threatening scowl that may have owed some of its intensity to my bothersome presence, but seemed more likely to have derived from inside him, for his gaze extended beyond me to some distant boundary. I sensed a wickedness about the man and his intentions, and I began to question the wisdom of following him into his office.

"Damn them anyway! The fools!" His startling outburst shattered my train of thought and stirred me to a position of readiness at the edge of my chair.

"Pardon me?"

"The bastards! They've destroyed the experiments -- the beautiful experiments. In their ignorance they've destroyed part of their own future, and their chance to control the world around them. Imbeciles, all of them!"

His pinkish cheeks glowed a feverish red as his excitement grew. I rose, somewhat in self-defense, for this unexpected frenzy of emotion seemed perhaps a precursor of even less rational behavior. He rose also, and with a guttural sigh that lifted the hint of alcohol across the desk, positioned his bulky form between my chair and a lengthy book cabinet swollen with oversized chemistry volumes. As a result, I was effectively blocked from the doorway.

"Do you know what they've done? Months of tedious, painstaking experimentation lost to a horde of malcontented fools! Do you have any idea what they've done?"

He paused as if to claim a moment's solace from me, but I could only stammer in confusion before the gritted, yellowing teeth that gave him the appearance of a salivating attack dog. His rumpled white beard contrasted sharply with the tan suit that girded him as stiffly as a corpse. As he advanced a step toward me, I watched his expanding eyes and tightly clenched hands with mounting apprehension.

"We had isolated the genes," he continued, with a note of despondency in his voice, "to build a working model of the self-replicating brain enhancement engine. Now we will have to repeat the countless hours of difficult work."

"Brain enhancement?" The idea struck me as so ludicrous that I couldn't avoid expressing my disbelief.

"Ah, such an elegant structure. A latticework of protein and phospholipids, molded exquisitely into the shape of electronic circuits by the tireless microbes and floating gently in a sea of saltwater. Of course, the latticework would serve no functional purpose by itself. We programmed amino acid sequences -- thankfully we've retained that, at least! Amino acid sequences to fashion helical protein structures that attach to the latticework and shape it to the proper dimensions. Just imagine what we've accomplished! Countless hours spent bending the proteins with proline groups, and linking the fragments with disulfide groups, in order to find just the right combination -- failure after failure after failure, and then, just as we near success, the mindless buffoons attempt to destroy it..."

He paused for a moment, perhaps to contemplate the greatness that had been stolen away from him by the infiltrators. I felt a breath of compassion building inside me, for suddenly the man seemed more pitiful than threatening.

"The very people who could most benefit from our work. The most intelligent students at the university, those with the potential for expanded intelligence. And they destroy their own futures!"

"But people can't be programmed..."

"Fool!" His violent reaction had me clinging to the back wall of the office, regretting my words as I hurriedly fashioned a possible escape route over his desk. "What do you think the brain is? Neural structures -- billions of them, together encoding an infinity of images and smells and sounds. Your entire being is just a result of how they grow and how they're conditioned and how they interact."

He paused again, and nodded his head in a rhythmic up-and-down motion: slowly at first, then suddenly faster and faster, until finally he burst out in a fit of raucous laughter that frightened me more than his spells of brooding silence.

"Ah, but they haven't damaged our amino acid sequencing files -- and now we've taken steps to safeguard them indefinitely! Once we reconstruct the bacterial colonies to manufacture the necessary proteins, we'll have most of our model back. And then the magnificent final step: programming the bio-device with charged molecules as storage bits. The result will be a knowledge device fashioned of nitrogen- and carbon-based molecules rather than semiconductor chips. Do you understand the implications? Perpetual reasoning devices manufactured in the billions by colonies of bacteria! Devices capable of symbiotic growth with the human brain! Tiny enough to be ingested and to replicate indefinitely in the brain's nutrient-rich environment! And placed discreetly" -- he smacked joyously at the thought -- "in the water bottles of those selected for the experiment. Ah, masterful!"

"In water bottles?" I felt convinced of his madness at that moment, but prudently nodded in agreement. The children and I had been drinking from water bottles. The man sat on the edge of the desk and thrust his outstretched hands into the air as a violent stream of curses issued forth from his driveling, jelly-red lips. His wrath was still directed at the vandals who devastated his life's work, but the distance between the transgressors and a more accessible bystander like me might not be sufficiently great. I considered breaking for the door, but hesitated as he spoke again.

"Of course, it will be necessary to deal with the masses. Poor wretches. They'll never understand the brain enhancers. There's no reason to try to expand their minds." He paused, then continued with a sinister laugh. "Their leaders must be dealt with, though...the rabble rousers, fomenters of violence, destined to be the victims of our indomitable policing technologies. Micro-drones -- like so many gnats in the air -- with just enough of an injectable poison, programmed with facial recognition software, to be released in the general vicinity of their targets, instructed to wait patiently as they remain charged by the sun, and finally to sweep in to their targets' heads to complete their task. Silent and unseen, unidentifiable and untraceable, and then self-destructing in the final act of their perfect mission."

His final remarks startled and sickened me, and impulsively I inched toward the open space created by his movement from the book cabinet to the desk. The sight and sound of the man could be tolerated no longer. Quickly but cautiously I stepped past him to the lab area, and then past the malodorous cluster of chemical sinks to the stairway. Without delay I descended the two flights to the basement -- utilizing perhaps half the steps in my rush -- and began a short passage through the cold, colorless concrete that shaped the uninviting tunnel area of the lower level.

Almost immediately signs directed me to the infirmary. I hurried around a corner to a glass door and pulled it open. Inside, a sizable lobby space containing medical and electronic equipment was separated from the infirmary by a glass wall, through which a stirring of numerous silhouettes was barely evident. In the center of the wall was a massive TV screen apparently used to view the patients in the larger room. I stepped up to the screen just as a young man clad in a green smock opened the lobby door and called in to me.

"I will be right with you."

"What?"

But before I could turn to him the TV screen exploded into a display of shapes and faces from beyond the glass wall. I was able to see an immense room, where a collection of shabby young men and women lined the walls in varying positions of resignation to their dismal surroundings. They appeared in such a state of degeneracy that my initial feelings of shock quickly gave way to disgust. Some were slumped sideways against the wall, their heads buried in the shadows created by their own bodies. Others curled in near-fetal positions on the floor, where only their compulsive quivering or the spittle oozing from their parted lips ensured their link to the living. Some grunted or giggled in sporadic outbursts of emotion. Most were silent. A few still moved about, although slowly, as if desensitized by long days of empty pacing that had blurred their unseeing eyes into waxy circles of fading light.

I noticed, as I forced myself to look away from the wretched scene in front of me, that a faint but intolerable mix of urine and vomit had permeated the lobby through the surrounding wall of glass. When I looked up again, the TV camera had begun scanning the infirmary. One expressionless face after another passed my inspection, each one revealing the unmistakable presence of a mind deadened by the invasion of unnatural substances. So unnerved was I by the relentless procession of faces that I barely noticed when a shadowy figure approached me from behind. I spun around, and then suddenly he was recognizable to me.

"Alan!

He grabbed my arm and hustled me along the wall of the infirmary and down a hallway that ended in an emergency exit, where upon release a piercing alarm was triggered, energizing our headlong retreat to the thickets surrounding Building 2, far beyond my earlier point of concealment. With the sun behind me -- dusk or dawn I was too addled to ascertain -- I was momentarily transfixed by a curiously prolonged squarish tube of a building reaching out to a wooded valley about 300 meters distant. Alan quickly smacked me back to awareness.

"Peter, we have to go to the main building." Babbage gave me directions.."

"Directions to where?"

"To get out of here."

He turned and began to forge a pathway through a small grove of decorative blue-green pine. We walked for 10, maybe 20, minutes, with my own state of mind alternating between skittish kitten and recalcitrant pack horse. Through bleary eyes there suddenly appeared a scene that stunned me to attentiveness: the majestic greystone cathedral-like anomaly that we had seen on our way into the country. To its side lay the newer institute structure, and behind them a patchwork of orchard and garden, a summertime pageantry of miniature fruit trees and pink and red roses that garnished a grouping of elaborate hedgerows.

I sank to the cooling earth and gawked at the unlikely panorama in front of me. The stillness was appealing, the quiet tranquilizing. I contemplated the irony of such an archaic edifice forming the backdrop to eye-blinking modern technology. Nothing in the moment made sense. I didn't feel ready to take the next step. I was overheated and grimy and thirsty and hungry, and a strange buzzing persisted inside my head. It was getting darker: either a few hours or an entire day had passed...the uncertainty seemed inappropriately humorous. Alan still waited. Droplets of sweat collapsed into my eyes and melted the towering buttresses above me into shapeless glittery matter. An unnatural silence, suggestive of watchful eyes behind the darkish windows, heightened the eerieness of the moment, so that I began to imagine an assembly of demons inside the building, until I was distracted by the sound of Alan's voice.

"I was looking all over for you. Some of the students said they saw you running through the halls."

"Yes, the children.."

"Children? I think they have summer programs going on."

"What happened to you, Alan?"

"I found Dr. Babbage...we were searching for you...they said you were in the infirmary...with the brain-expansion patients."

I remained quiet, unsure of my ability to accurately recall any of my recent experiences.

"Babbage told me the best way out -- Peter, are you listening?"

"Yes. I'm sorry."

"We should go."

We cautiously approached the entrance to the stately old main
building of the Logic Institute. No students or staff members
interrupted our passage through the churchlike lobby that seemed,
amidst air heavily scented by the aged woodwork, somehow
violated by our presence. Upon hurriedly reaching a back exit, we
were released into the nature preserve we had seen a few minutes
before, an overgrowth of colorful barberry and burning bush and
wild grass and roses and fruit trees between the old and new
buildings and nestled behind the ominous hedge walls, all of the
foliage providing a pleasantly welcome semi-secluded makeshift
walkway along the newer structure's 300-meter wall, toward the
river and a densely forested row of foothills. We emerged from the
botanic splendor and began to follow the edge of the river that
would lead us through the woods and back home. As I turned to
look back for a final time, I saw the sun reflecting off a set of
silvery bars that enclosed a doorway on the top floor of the end of
the Logic Institute.

Home

In retrospect, it's far from inexplicable
that we should be entwined in this umbilical
connection with an influential Institute
with links to technomania so absolute.

But after being party to such physical
and mental stress, it struck us as remarkable
that realms of inspiration and vitality
could coexist with worlds of surreality.

Thanks to the provisions and directions from Dr. Babbage, we made it back through the mountains in less than two days. The dry, cool hilltop air contributed to a quick recovery from any mind-altering substance that may have rendered me physically listless and mentally overreactive – both explanations, possibly, for any distorted perceptions that may have intruded on my sense of reality. Alan seemed to have experienced no ill effects from our transcendental journey to the sphere of our higher education.

The knapsack of provisions given to us by Dr. Babbage contained a letter, unexpected but gratefully received. The contents were as follows:

Alan and Peter:

I'm greatly troubled by recent events in my country, and moreso by the philosophy, developed here over several decades, that unfettered individual success is the key to a vibrant society. It seems to have justified a winner-take-all mentality. The myth is that everyone will ultimately be swept into the prosperity of a few. In truth, it's the other way around: a society in which all members are supported provides many more individuals the opportunities to express themselves and to develop their talents.

Here in Bin, we once seemed to be on the proper path to that goal. A guaranteed annual stipend has unleashed the latent genius of a generation of young people who are no longer indebted, no longer forced to work low-wage jobs to survive, no longer deprived of the freedom to explore and cultivate their dreams and desires. The arts and sciences have thrived. But that's not enough to transform a society in a beneficial way. Without the ability of people to regulate their own excesses, narcissistic personalities take over, exploiting the system, bullying the weak, extracting the wealth.

And, as we've seen in Bin, begetting monsters in the lab -- both the AI lab and the bio lab. Unregulated flows of research money have elevated the self-serving experts to unimaginable positions of ascendancy. Business people control the flow of news; robots make policing decisions; drones use facial recognition to monitor citizens; gaming enthusiasts blur the line between reality and imagination; biochemists conduct furtive experiments on the unsuspecting masses.

Popular resistance has been thwarted by tech-savvy billionaires who started with benevolent idealism but quickly succumbed to the lure of power. The two of you may have seen hints of that resistance in the few hours you were here. Reformist media is beginning to take the story beyond our borders. Perhaps the three of us will meet again soon, after my countrypeople finally see the dangers in an individualist mentality. Ha! The irony! Tyranny is not in cooperation, but rather in conquest and condescension. Bin leaders seemed to recognize this not long ago. But we squandered the opportunity to make it last. I hope your own nation doesn't falter as we did.

With heartfelt wishes for your safety and happiness,
Henry Babbage

In the days and weeks to follow, we gradually heard more about Bin's AI revolution, practical and fanciful, conventional and bizarre, rewarding and ruinous. How regrettable it was that our short stay in the country afforded us no entry into the vibrant and productive society that had appealed to our inquisitive minds! Instead we experienced a sampling of the alternative Bin universe.

Various international news reports have told of the dramatic social and political changes in the country, including unsubstantiated tales of violent pushback against some of the more controversial cognitive studies. Memories of Golem's lab occasionally flashed back into my consciousness.

But most of the mainstream reporting simply repeated what I had known before. And it generally confirmed the appraisal of a country in the throes of too-rapid technological change. Jobs have continued to disappear in Bin, and despite comforting references to the job-creating aftermath of the long-ago Industrial Revolution (which actually took 60 years to develop), Bin had little foreknowledge of the eventual means of employment for millions of human beings.

So Bin had turned to a full-scale guaranteed income program. With much of the pressure of job-seeking removed, occupations in Bin have taken on highly creative forms, as many once-frustrated artists and actors and musicians, assured of some minimum financial security, have become free to help sustain the cultural needs of the nation. Some have pursued humanitarian interests. Others work incessantly to implement a social-oriented political system. Far removed from all this were the many liberated young Binians -- as Alan and I well knew -- who had turned to new forms of technology, especially extreme AI. Game-playing enthusiasts translated their chimeral instincts into real-life challenges for anyone willing to test his or her cognitive skills. Alan and I were accidental participants. I frequently recall the steely sliding wall of the human snare in the mountains; the recursive sequence of delirium-inducing waiting rooms; the maniacal drive through the garden labyrinth as we tried to enter the institute. And the children's games -- what were their purpose? Perhaps even the respected educators of this AI-crazed nation were convinced that the future depends on the shaping of little minds in parallel with the workings of the great machinery that was beginning to control society.

Babbage told us about the well-positioned authoritarians who place individual interests over the needs of society. But there's an ongoing clash between good and evil in Bin, as the nation's guaranteed income has also inspired numerous business cooperatives, especially in clean energy, which has created high-level positions in technology as well as labor-intensive jobs requiring less education. Dr. Babbage was most enthusiastic about this. The sun, he once informed us, supplies thousands of times more energy than people could ever use. The greatest imaginable energy source is out there for the taking by the nation smart enough to take it.

Bin and its partners, once among the world's biggest polluters, are constructing a modern "Great Wall," a Global Energy Network that will connect the whole world with a wind farm at the North Pole and solar stations at the equator, in a manner that will demand cooperation rather than competition among nations. For all humans, then, an opportunity for peacemaking as well as for profit. Nations will be crisscrossed by solar roads that power the homes and businesses lining their edges and the cars on their surfaces. Great battery storage projects are capturing excess energy, each with the capacity to power tens of thousands of homes year-round. A fortuitous consequence is that the once-maligned word 'social' has taken on a more favorable connotation as nations are forced to collaborate on energy solutions.

How does Bin pay for it all? The guaranteed income has stimulated considerable economic return. In addition, since many of the benefits accruing to big business have derived from automation, and thus from decades of public input, taxpayer funding, and government research, the beneficiaries of the largesse had to finally face the realization that some payback is reasonable and proper, and logical. Their contributions to infrastructure and education have significantly increased.

~~~~~

Today life is nearly back to normal for me. But my thoughts often return to the alternative Bin world. At times I think of little Addie, and I can't help but smile. Little else of my memories of the Logic Institute seems real. The engineering deceptions; the outrageous children's games; the dementia of the adults; the human degradation surrounding brain-enhancing lab experiments. Alan saw much less of it than I. I seem to have been affected more by it. But as time passes I lose some confidence in my recollections.

Although we regularly hear new rumors about developments in Bin, both virtuous and vilifying, the nation remains largely isolationist. Alan and I received a second, research-related, message from Dr. Babbage, who has apparently avoided the controversy over brain science with his more mainstream work in language and cognition theory.

I stare blankly at a window blackened by some inconsequential hour of night, and I picture the faces of the children, recalling their enthusiasm for the logic games, wondering where they might be. Of course they were real. I remember what Dootil said about a person's individual realities. Dootil was such a foolish little man.

Author

Paul Buchheit has been writing for progressive journals for many years. His recent book "Disposable Americans" was published in 2017 by Routledge. He was named one of the 300 Living Peace and Justice Leaders by the TRANSCEND Network for Peace Development.

Mr. Buchheit's doctorate is in Computer Science (U of IL, 1991), and several years of subsequent research focused on Cognitive Science and theories of language development.

To correspond with the author, send email to
paul@youdeservefacts.org

Image Credits
(All in the U.S. Public Domain)

Cover
 The Garden of Eden with the Creation of Eve. 1630s painting by Jan Brueghel the Younger
 The eighth circle of hell. Gustave Doré, 1870

What We Learned
 Pxhere: CC0 Public Domain, Free for personal and commercial use, No attribution required

Quest I, Quest II, Quest III
 Pixabay: CC0 Creative Commons, Free for commercial use, No attribution required

What I Dreamed
 William Blake (1757–1827), The Stygian Lake, with the Ireful Sinners Fighting

 Gustave Doré's illustration to Dante's Inferno. Plate XIII: Canto V: Minos judges the sinners.

 Gustave Doré's illustrations to Dante's Inferno, Canto 28

 Gustave Doré's illustrations to Dante's Inferno, Plate LXV: Canto XXXI: The titans and giants.
 Gustave Doré's illustration to Dante's Inferno. Plate IX: Canto III: Arrival of Charon.

 Gustave Doré's illustration to Dante's Inferno. Plate XIV: Canto V: The infernal hurricane that never rests..

 Gustave Doré: Dante's guide in Inferno Canto 21 between ditches five and six in the eight circle.

Gustave Doré: Punishment of the Simonists

William Blake (1757–1827), Antaeus setting down Dante and Virgil in the Last Circle of Hell

Quest IV
Adapted from Erik Werenskiold (1855–1938), Children playing (1891)

Quest V, Quest VI
Pixabay: CC0 Creative Commons, Free for commercial use, No attribution required

Quest VI
Author's creation of neural network

Quest VII
Adapted from Children and their teacher at school in Chelsea, England, during a spelling lesson, 1912

Quest VIII
Pxhere: CC0 Public Domain, Free for personal and commercial use, No attribution required

Quest IX
Adapted from Poster for a theatrical adaptation of Strange Case of Dr Jekyll and Mr Hyde, 1880s